Ships of War

The Battle of Trafalgar

Ships of War
The Development of Warships by the Navies of the World During the Later 19th Century

Edward J. Reed
Edward Simpson &
J. D. Jerrold Kelley

Ships of War: the Development of Warships by the Navies of the World During the Later 19th Century
by Edward J. Reed, Edward Simpson & J. D. Jerrold Kelley

Leonaur is an imprint of
Oakpast Ltd

Material original to this edition and presentation of the
text and illustrations in this form
copyright © 2012 Oakpast Ltd

ISBN: 978-0-85706-954-2 (hardcover)
ISBN: 978-0-85706-955-9 (softcover)

http://www.leonaur.com

Publisher's Note

The views expressed in this book are not necessarily
those of the publisher.

Contents

Preface	7
Introduction	9
The British Navy	21
The French Navy	81
The Italian, Russian, German, Austrian and Turkish Navies	116
The United States Navy in Transition	156
United States Naval Artillery	207
Ships of the Minor Navies	256

Appendices

Submarine Warfare	265
The Question of Type	287
Range of Guns	317

Preface

After many years of neglect, the people of this country have awakened to the necessity of creating a modern fleet. Proud as they were of the navy's achievements in the past, they failed for a long time to exhibit any interest in its present or future, and met all claims for its re-establishment by a denial of its usefulness, or by a lazy optimism of indifference which smilingly put the question by. Indeed, at one time, the popular solicitude disappeared completely, and outside of the service there was manifested neither an alarm at its degeneracy nor an appreciation of the dangers this made possible:

> With an apathy inexplicable upon any rational grounds, the notes of warning sounded by experts were unheeded, and the law-makers contented themselves by pinning their faith to what they called "the creative possibilities of American genius.

They accepted this fallacy as a fact, they made this phrase a fetish, and with a fatuous hope believed it could, by some occult inspiration, in the event of sudden, sharp, and short war, save them from the fighting-machines which twenty years of tireless experiment had perfected abroad. In the end, by a neatly balanced policy of pride and folly, the navy was exhausted almost to dissolution. Then Congress lazily bestirred itself to action, and prescribed as a remedy three unarmoured cruisers and a despatch-boat.

Heroic treatment, not homoeopathy, was needed; but, thanks to a naturally vigorous constitution, the bolus sufficed to lift the patient out of the throes, and to encourage him into a languid convalescence. Luckily, the vessels became a party question, and their historic tribulations did so much towards educating the nation that a public sentiment was aroused which made a modern navy possible. It must be confessed, however, that the demand even yet is not so vociferous as to dominate all. Other issues, though there is apparent everywhere a quickening desire for the country to take, if not the first, at least a respectable, place among the great maritime powers.

With these new ideas came a desire for information which could not be satisfied, because, curiously enough, the popular literature of the subject is meagre, or rather it is unavailable. There are treatises in plenty which soar beyond the skies of any but experts; there are handy manuals wherein the navy, like the banjo, is made easy in ten lessons; but between these extremes nothing exists which is accurate, and at the same time free from those dismal figures and dry-as-dust facts that are so apt to discourage a reader at the outset.

To meet this want, which was one by no means "long felt," these articles were originally published in *Harper's Magazine*, and with a success that seemed to justify their collection in a more available, if not a more permanent, form. It may be said now that no changes of any moment have been made in the text, that the notes attempt only to bring down the data to the latest date, and that the appendices are needful additions which the limited space of a monthly publication necessarily forbade. The reader who has not followed the progress of naval war construction will undoubtedly find many surprises, both in achievement and promise, which may be difficult to understand, yet it is hoped that the non-technical manner in which Sir Edward Reed and Rear-admiral Simpson have written will do much to make plain this important national question. Both these gentlemen are authorities of the first rank, both are luminous writers, and each in his own country and own sphere has had an important influence upon warship design and armament. To those who read within the lines there awaits a mortifying realization of our inferiority; for during all the years that this country—masterful beyond compare in other material struggles—was so successfully neglecting its navy, foreign designers were achieving triumphs which are marvellous. With this knowledge there is sure to come a high appreciation of the intelligence exercised; for the evolution of the battleship has been so rapid, and the resultant type has so little in common with the wooden vessel of our war, that those who have solved the problems have practically created a new science.

Introduction

During the last thirty years the changes in naval science have been so much greater than in its whole previous history as to be epoch-making. Between the wooden vessel of 1857 and the metal machine of 1887 there exist in common only the essential principles that each is a water-borne structure, armed with guns and propelled by steam. Beyond this everything is changed—model, material, machinery, rig, armament, equipment. In truth, so radical are the differences, and so sudden have been the developments, that authorities are widely separated in opinion, even upon such a primary question as a universally accepted system of classification. But as this is necessary to a proper appreciation of the subject, a generalization may be made in which war-vessels are divided into armoured and unarmoured types, the former including battleships, and the latter those employed in the police of the seas, in commerce protection or destruction, or in the attack of positions which are defenceless.

In the absence of any accepted differentiations of these classes, the new British nomenclature may be adopted with safety, for to a certain degree it explains the terms and includes the types now used so variously in different navies. Under this armoured vessels are grouped into (1) battleships, (2) cruisers, (3) special types, such as rams and torpedo-boats, and (4) coast-service ships; and unarmoured vessels comprise (1) cruisers, (2) sloops, (3) gun-vessels, (4) gun-boats, (5) despatch-vessels, and (6) torpedo-vessels.

> As it was impossible to unite all, the qualities which are to be desired in a ship-of-war in a single vessel, it became necessary to divide the leading types into subdivisions, each specially adapted to the use of a particular arm, or to perform some special service. For the battleships designed for naval opera-

tions in European waters great offensive and defensive powers and evolutionary qualities are essential, while the highest sea-going qualities, including habitability, are, in the opinion of some, less essential. For sea-going battleships offensive and defensive strength must be partially sacrificed in order to secure unquestionable sea-worthiness. In ocean-going battleships canvas is a valuable , auxiliary. In battleships for European waters, masts and yards involve a useless sacrifice of fighting power. . . . Heavily armoured ships intended for the line of battle must necessarily carry powerful guns. They must be able to traverse great distances, and must therefore have considerable storage for coal. Great speed is required to enable them to meet the inevitable contingencies of an engagement. In a word, the class of ships which may be called battery-ships must be furnished with very considerable offensive and defensive power. Their great size, however, and the enormous weight of their armour and armament, necessitate such displacements as render them unfit for coast defence. (Brassey)

While the antagonistic elements of offence, defence, speed, or endurance have caused the main differences of design in all types, the greatest variances with battleships are found in the distribution of armour for protection. A hasty summarization of the policies now adopted by the great maritime nations shows that the French generally adhere to a complete armour-belt at the water-line, that the Italians have in their latest ships totally abandoned side-armour, and that the English favour its partial employment. The popular idea that armour consists only of thick slabs of wrought-iron or steel, or of steel-faced iron, bolted to a ship's side, is erroneous.

In the earlier broadside ships, this view was practically correct; they had no armour or protected decks, the decks being covered only by thin plates fitted for structural purposes. But in the *Devastation* class, and all subsequent ships, considerable and increasing weights of material are worked into the deck armour, and with good reason. Experiments showed conclusively that horizontal protection at the top of the armour-belt, or citadel, was of vital necessity, and even now (1887) it is open to question whether the provision made for horizontal protection in relation to vertical armour is as large as it might advantageously be. (W. H. White)

The factors which have most influenced the problem are the torpedo, ram, and gun. Of these the last is indubitably of the highest importance, for the number and nature, the effective handling, the disposition and command, and the relative protection of the guns are the elements which control most powerfully the principles of ship design. In the first stage of the contest between gun and armour the defence was victorious, but so rapidly have the art and science of ordnance developed that today the power of the heaviest pieces as compared with the resistance of the heaviest armour is greater than ever before.

The story of the contest can be briefly told. In 1858 the armament of the newest ships was principally a broadside battery of 32-pounders; in this were included a few 56-pound shell guns and one or two eight-inch 68-pounders, though of the whole number not one had an energy, that is, a force of blow when striking, sufficient to penetrate four and a half inches of wrought-iron at short range. In the earliest ironclads—the French *La Gloire* and the English *Warrior*—batteries mainly of nine-inch calibre were carried, the latter mounting forty guns of all kinds. The *Minotaur*, the first representative of the next English type, had fifty guns, but after this class was launched there appeared that distinctively modern tendency to decrease the number of pieces while increasing the intensity of their fire. The succeeding vessels carried from fourteen to twelve pieces, until, in 1871, the principle of concentration reached its maximum in giving the *Inflexible* only four guns. These, like the *Warrior*'s, were muzzle-loaders, and their relative dimensions and power may be compared as follows:

	Warrior.	Inflexible.
Weight of gun	4¾ tons.	80 tons.
Length	10 feet.	26 feet 9 inches.
Calibre	8 inches.	16 inches.
Powder charge	16 pounds.	450 pounds.
Weight of projectile	68 "	1700 "
Energy at 1000 yards	452 foot-tons.	26,370 foot-tons.
Penetration of 4½ inches of wrought-iron at short range	None.	....
Penetration of wrought-iron at 1000 yards	....	23 inches.

The term energy, when employed to indicate the work that a gun can perform, is expressed in foot-tons, and signifies that the amount developed is sufficient to raise the given weight in tons to the height of one foot. The piercing power of the *Inflexible*'s projectile was, under the same conditions of charge and range, sufficient to penetrate twenty-five feet of granite and concrete masonry, or thirty-two feet of the best Portland cement.

When the thickness of armour-plating increased, gun-makers tried to overcome the resistance by giving greater energy to the shot. As this required large charges of powder and very long guns, muzzle-loaders became impracticable on shipboard, and were supplanted by breech-loaders. From this stage guns developed greatly in power until, in 1882, those designed for the *Benbow* were to weigh 110 tons, to be 43 feet long and 16¾ inches in calibre, and with 900 pounds of powder and an 1800-pound projectile were to develop 54,000 foot-tons, or an energy sufficient to penetrate thirty-five inches of unbacked wrought-iron at one thousand yards. The guns for the latest English ships, the *Trafalgar* and the *Nile*, weigh 67 tons, are 36 feet 1 inch in length and 13½ inches in calibre, and with a 520-pound charge and a 1250-pound projectile are expected to develop 29,500 foot-tons, or an energy sufficient to penetrate an iron target twenty-two and a half inches thick at a thousand yards. These results apparently show a retrogression in power, but a comparison of the *Inflexible*'s and *Trafalgar*'s batteries proves that the more modern gun of the latter weighs 13 tons less, is 2½ inches smaller in calibre, fires a shot 450 pounds lighter, and yet develops an energy greater by 3000 foot-tons.

This gain is mainly due to the improvements made with powder and projectiles. In 1883 a 403-pound Whitworth steel shell penetrated a wrought-iron target eighteen inches thick backed by thirty-seven inches of well-packed wet sand, one and a half inches of steel, various balks of timber, and sixteen feet more of sand. When the projectile was recovered after this stratified flight it was found to be practically uninjured. On the Continent, where breech-loaders were favoured earlier than in England or with ourselves, the heaviest rifles afloat are the 75-ton, 16.54-inch calibre, French, and the 106 tons, 17-inch Italian guns. These are, however, not the largest pieces designed, for there is an 120-ton Krupp gun, and the French have projected one which will weigh 124 tons, be 18.11 inches in calibre, and fire a 2465-pound projectile—over a ton—with a powder charge of 575 pounds. A comparison of the Krupp 120-ton gun with the 110-ton Armstrong shows that the former is more powerful; that its projectile is much heavier, and the initial velocity and pressure are smaller. The results at the recent test were as follows:

	Armstrong.	Krupp.
Charge	850 pounds	847 pounds.
Shot	1800 "	2315 "
Velocity	2150 feet	1900 feet.
Pressure	19.9 tons	18.8 tons.
Energy	57,679 foot-tons	67,928 foot-tons.

From the *Warrior* to the *Inflexible* the evolution of design was based upon a principle that sought the best results for the offence in small, powerful batteries, with all-around fire and armour protection; and for the defence, in thick armour carried over the vitals of the ship. This was satisfied by larger weights of armour and a smaller ratio of armoured part to total surface. Wrought-iron armour was also replaced by compound, with a corresponding gain of twenty per cent, for equal thicknesses, and at present all steel plates, of which great things may reasonably be expected, are now employed by France and Italy. In 1861 the *Minotaur* was belted throughout her 400 feet of length with 1780 tons of armour, or with a weight nearly double that given to the *Warrior* two years before. The *Inflexible* has 3280 tons, and the *Trafalgar* 4230 tons, of which 1040 are fitted horizontally. The maximum thickness of the *Warrior*'s wrought-iron armour is 4½-inches, of the *Devastation*'s, 12 inches, and of the *Inflexible*'s 24 inches; the compound (iron steel-faced) armour of the *Trafalgar* is 18 and 20 inches thick, and the *Baudin* and *Formidable* have 21.7 inches in solid plates of steel. These, of course, are some of the dry-as-dust figures before referred to, and they are cited only to assist a comparison, their mere enumeration having no scientific value, because the disposition and character of the plates are unconsidered.

To meet this development of offence and defence many changes in design have been adopted. The *broadside system* of the first armoured ships was followed in 1863-1867 by a *belt and battery type*, wherein the principal guns, much reduced in number, were carried in a box battery amidships, and given a fore-and-aft fire by means of recessed ports or outlying batteries. In 1869 the Admiralty adopted the *breastwork monitor*, a low free-boarded structure, which was plated from stem to stern in the region of the water-line, and had in its central portion an armoured breastwork that carried at each end a revolving turret. In 1870 this type was pronounced unsafe, and after a careful investigation by a special committee on design certain modifications were recommended. These did not affect materially the essential features of Sir Edward Reed's plan, for the complete water-line belt and the central armoured battery were retained; and today many of the critics who then denounced it claim that, after all, it is the true type of an ideal battleship.

In 1872 the Italian naval authorities accepted the conclusions of the British committee, and laid down the first *central citadel* battleships, now known as the *Duilio* and *Dandolo*; and about the same time Mr. Barnaby, the new chief constructor of the British navy, brought forward a similar design in the *Inflexible*. The engines, boilers, and the bases of two turrets in this vessel are protected by an armoured box-shaped citadel, from the extremities of which a horizontal armoured deck extends fore and aft below the water-line; above this deck an armoured superstructure completes the free-board, and has its unprotected spaces at the water-line, subdivided into numerous water-tight compartments. This ship met with so much hostile criticism that a committee was appointed to investigate the charges, but in the end the Admiralty plans were officially sustained.

The French, with characteristic ability and independence, have in the meantime made many notable departures from their first types of broadside ships. Believing in the association of heavy guns with light ones—mixed armaments, as they are called—the central armoured case-mate of wholly protected guns has been rejected in order to give a maximum thickness of plating at the water-line. The largest guns are mounted *en barbette*—that is, in towers which protect the gun mechanism, and permit the pieces to be fired, not through port-holes, but over the rim of armoured parapets. The French constructors reason, and with justice, that no single shot from a heavy gun should be wasted, and that, in addition to an extended range, gun captains must be enabled, by keeping their eyes upon the enemy, to select the best opportunity for firing. With broadside pieces this is impossible, for apart from the limited range, and the obscurity caused by the smoke, the port-holes through which the sighting has necessarily to be done are almost choked by the gun-muzzles. Turrets have their objections also, because the poisonous gases which formerly escaped wholly from the muzzle will, as soon as the breech is opened, rush into the turret and make it almost uninhabitable. Often after one discharge the air becomes stifling, and in the *Duilio* it deteriorated so quickly as to be unfit for respiration until a part of the turret-roof had been lifted. Then, again, structural difficulties not easily overcome in the turret are simplified in the barbette, as the latter, with equal gun facilities, weighs fifty per cent, less, and at the same time escapes all those chances of disablement which a well-placed shot is almost sure to cause in any revolving-system. At sea the chance of hitting the gun is never great, and the main things to protect are the gun machinery and the gunners; the armoured wall of the barbette tower does this for the former, and the latter have a fair fighting chance afforded by the gun-shield. Of course war is not deer-stalking, and the patriot who wants to go into battle so fully protected as to be in no danger had better stop playing sailor or soldier, and take to the woods before the fighting begins. In addition to the heavy ordnance, the French mount a number of lighter pieces, and carry powerful secondary batteries of rapid fire and machine-guns; and sufficient armour defence is given by a belt at the water-line, an armoured deck, and a glacis and parapet for the barbette. It is quite probable that these purely military terms may seem odd when applied to ships, but they are the only ones which can exactly explain what is meant, and, after all, they show how much a battleship has become a floating, transferable fortress.

The Italians were not altogether satisfied with the *Duilio*, as she lacked the high speed and coal endurance which they deem essential in any Mediterranean naval policy; so in 1878 they adopted an idea advanced some years before in England, and startled the world with the *Italia* type. In this ship protection is given, not by vertical or side armour, but by an armoured deck, between which and the deck above there is a very minute subdivision of the water-line space. The system is based upon the theory that the power to float must be obtained, not by keeping our projectiles, but by so localizing their effect as to make any penetration practically harmless. The *Italia*'s heavy guns are carried in a central armoured redoubt, at a height of thirty-three feet above the water-line, and with their machinery and fittings weigh over two thousand tons. This fact shows the magnitude of the task accepted by her designer, for it means that a load nearly equal to the total weight of a first-class line-of-battle ship of the last century has to be sustained at this great elevation. Besides the main battery of four 106-ton guns, eighteen six-inch breech-loading guns are carried—twelve in broadside on the upper battery deck, four in the superstructure before and abaft the redoubt, and two under cover at the extremities of the spar-deck. The redoubt is protected by seventeen inches of compound iron, inclined twenty-four degrees from the vertical; and the complete armoured deck, which is nearly three inches thick, dips forward to strengthen the ram, curves aft to cover the steering gear, and, at the ship's sides, extends six feet below the water-line.

To cope with this formidable rival, who, whether right or wrong in principle, must, under England's policy, be surpassed, the ships of the *Admiral* class were designed. In these the main battery is mounted in two barbettes built high out of water, near the extremities of the vessel, while in a central broadside are carried the armour-piercing and rapid-fire guns. The engines, boilers, and barbette communications are protected by a water-line belt of thick armour which covers about forty-five *per cent*, of the ship's length; at the upper edge of this there is a protective deck, and at its ends athwart ships bulkheads are erected; before and abaft the belt and beneath the water-line there is a protective deck, together with the usual minute subdivision into water-tight compartments. The barbettes and the cylindrical ammunition tubes which extend from the belt-deck to the barbette floors are strongly armoured. Owing to the strong protest made against these vessels, more efficient armour protection has been given to the battle-ships lately laid down.

From this very hasty and incomplete review it may be gathered that the first and most lasting influence in the development of battle-ships is due to France and England, though the *Monitor* had no little share in the result. It is difficult to say, in the ceaseless struggle for something which, if not good, is new, what may be the outcome of the latest efforts to revolutionize the question, or, curiously enough, to bring it back to the point whence its departure was taken. Whatever may be the courage of one's opinion, there is not sufficient data—a first-class war can only supply these—upon which to say, yea, yea! Or nay, nay! And prophecy is certain to be without honour, especially as the discussions given in the appendices demonstrate how the wisest and most experienced have no substantial agreement in views. An editorial in a late number of the *Broad Arrow* declares that:

> The days of armoured plate protection are, in the opinion of many thinking men, coming to a close. The gun is victorious all along the fine, and the increased speed given to the torpedo-boat, taken in conjunction with the destructive efficiency attained by the torpedo, makes it a questionable policy to spend such large sums of money as heretofore upon individual ships.

There is no room here to give the various arguments, though very clever and ingenious they are, by which this position is fortified; it may be added, however, that to a large degree this is the opinion of Admiral Aube, the late French minister of marine, and undoubtedly this declaration re-echoes the shibboleth of those other French officers who, in the absolute formula of their chief, Gabriel Charmes, insist that "a squadron attacked at night by torpedo-boats is a squadron lost."

English authorities, with a few notable exceptions, do not go so far as their more impulsive, or, from the Gallic standpoint, less conservative neighbours. chief constructor White believes that at no time in the war between gun and armour has the former, as the principal fighting factor, so many chances of success. He concedes the value of light, quick-firing guns in association with heavy armaments, grants the importance of rams, torpedoes, submarine boats, and torpedo-vessels generally, but denies that the days of heavily armoured battleships are ended. Lord Charles Beresford asserts that the value of large guns at sea is overestimated, advocates from motives of morals and efficiency mixed armaments, agrees to the great, yet subordinate, importance of the usual auxiliaries, and insists that England builds cumbersome and expensive battleships only because of their possession by her dangerous rivals.

There are equally rigorous disagreements upon all the other types of armoured, unarmoured, and auxiliary vessels, as needs must be, so long as the naval policies of no two nations can be alike. England and Russia are at opposite poles, so far as their environments are concerned, and between France and Turkey the differences are as radical as their national instincts and ambitions. But, among all, England is as isolated as her geographical situation. Whatever fleets other nations may assemble, whatever types other countries may deem best for their interests, England, whose existence depends upon her naval strength, must have all; not only the best in quality, but so many of every class that she will be able to defend her integrity against any foe that assails it. England can take no chances.

Upon one point alone, the necessity of high speed, is there substantial agreement. Less than four years ago fifteen or sixteen knots were accepted as a maximum beyond which profitable design could not be urged. Greater speed, it is true, had been attained by our first type of commerce destroyer. In February, 1868, the *Wampanoag* ran at the rate of 16.6 knots for thirty-eight hours, and made a maximum of 17.75 knots; but great as was the achievement, there is a general acceptance of the fact that this vessel was a racing-machine, and not in the modern sense a man-of-war.

Fighting-ships, with the power to steam thousands of miles at sea without recoaling, are now being built under contracts which, for every deficiency in speed or horse-power, pay penalties that at our former summit of expectations would have been prohibitive to ship construction; and, what is more startling yet, the bonus which goes to any increase upon this speed proves the co-relation between scientific attainment and popular appreciation of the subject, and shows how readily the impossibilities of yesterday become the axioms of tomorrow.

The development of speed has therefore a special interest. Between 1859 and 1875, that tentative period which led to such wonderful realizations, the highest speed, under the most favourable circumstances, of large war vessels was fourteen knots; in the smaller classes of unarmoured ships it ranged between eight and thirteen, while that attained by fast cruisers was from fifteen to sixteen and a half knots. In 1886 Italian armoured vessels made eighteen knots. Cruisers like the Japanese *Naniwa Kan* and the Italian *Angelo Emo* reached nearly nineteen, and the *Reina Regente*, launched in February last, is expected to steam over twenty. Torpedo vessels beginning in 1873 with fourteen knots are now running twenty-five, and at the same time the type has so much increased in size and importance as to be an essential and not an accessory in naval warfare.

It is impossible to explain the difficulties which have beset this development, because the conditions that surround any attempt at speed-increase are such as can be properly understood only by those who have technical training; and then, too, the great ocean racers have so much accustomed the public to wonderful sea performances that the results are accepted without a knowledge of the credit which is due the mechanical and marine engineers who have achieved them. But with greater experience the higher, surely, will be the appreciation which everyone must give; for, in the words of chief constructor White:

> When it is realized that a vessel weighing ten thousand tons can be propelled over a distance of nine knots in an hour by the combustion of less than one ton of coal—the ten-thousandth part of her own weight—it will be admitted that the result is marvellous,.... the way of a ship in the midst of the sea is beyond full comprehension.

It is often asked which has the better fleet, France or England. Who can tell? No one definitely. Admiral Sir R, Spencer Robinson, late comptroller of the British navy, declares, in the *Contemporary Review* of February, 1887, that:

> The number of armoured vessels of the two countries may be stated approximately as fifty-five for England and fifty-one for France. Without going into further details, taking everything into consideration, giving due weight to all the circumstances which affect the comparison, and assuming that the designs of the naval constructors on each side of the Channel will fairly fulfil the intentions of each administration (a matter of interminable dispute, and which nothing but an experiment carried to destruction can settle), the ironclad force of England is, on the whole, rather superior to that of France. A combination of the navy of that power with any other would completely reverse the position. I should state as my opinion, leaving others to

judge what it may be worth, that in fighting power the unarmoured ships of England are decidedly superior to those of our rival's; but if the *raison d'être* of the French navy is—as has been frequently stated in that country, and by none more powerfully and categorically than by the French minister of marine—the widespread, thorough destruction of British commerce, and the pitiless and remorseless ransom of every undefended and accessible town in the British dominions, regardless of any sentimentalities or such rubbish as the laws of war and the usages of civilized nations; and if at least one of the *raisons d'être* of the British navy is to defeat those benevolent intentions, and to defend that commerce on which depends our national existence and imperial greatness—then I fear that perhaps they have prepared to realize their purpose of remorseless destruction rather better than we have ours of successful preservation.

A long sentence this, but it emphasizes the great axiom that war is business, not sentiment, and teaches a lesson which this country will do well to learn. Fortunately, we are at last out of the shallows, if not fairly in the full flooding channel-way, though many things are yet wanting with us. Perhaps this over-long chapter cannot be made to end more usefully than by quoting in proof of this the concluding paragraph of that brilliant article on naval policy which Professor James Russell Soley, United States navy, contributed to the February (1887) number of *Scribner's Magazine*:

> It is the part of wisdom to study the lessons of the past, and to learn what we may from the successes or the failures of our fathers. The history of the last war is full of these lessons, and at no time since its close has the navy been in a condition so favourable for their application. At least their meaning cannot fail to be understood. They show clearly that if we would have a navy fitted to carry on war, we must give some recognition to officers on the ground of merit, either by the advancement of the best, or, what amounts to nearly the same thing, by the elimination of the least deserving; that we must give them a real training for war in modern ships and with modern weapons; that the direction of the naval establishment, in so far as it has naval direction, must be given unity of purpose, and the purpose to which it must be directed is fighting efficiency; that a naval reserve of men and of vessels must be organized capable of mobilization whenever a call shall be made; and, finally, that

a dozen or a score of new ships will not make a navy, but that the process of renewal must go on until the whole fleet is in some degree fitted to stand the trial of modern war. Until this rehabilitation can be accomplished the navy will only serve the purpose of a butt for the press and a foot-ball for political parties and its officers—a body of men whose intelligence and devotion would be equal to any trust will be condemned to fritter away their fives in a senseless parody of their profession.

The British Navy
by Sir Edward J. Reed

When timber gave place to iron and steel in the construction of warships, the naval possibilities of Great Britain became practically illimitable. Prior to that great change the British Admiralty, after exhausting its home supplies of oak, had to seek in the forests of Italy and of remote countries those hard, curved, twisted, and stalwart trees which alone sufficed for the massive framework of its line-of-battle ships. How recently it has escaped from this necessity may be inferred from the fact that the present writer, on taking office at the Admiralty in 1863, found her majesty's dockyards largely stored with recent deliveries of Italian and other oak timber of this description.

And here it may not be inappropriate for one whose earliest professional studies were devoted to the construction of wooden ships, but whose personal labours have been most largely devoted to the iron era, to pay a passing tribute of respect to the constructive genius of those great builders in wood who designed the stanch and towering battleships of the good old times. Skilful, indeed, was the art, sound, indeed, was the science, which enabled them to shape, assemble, and combine thousands of timbers and planks into the *Grace de Dieu* of Great Harry's day (1514), the *Sovraigne of the Seas* of Charles's reign (1637), the *Royal William* of half a century later (1682-'92), the *Victory*, immortalized by Nelson, and in our own early day such superb ships as the *Queen*, the *Howe*, and scores of others. Only those who have made a study of the history of sea architecture can realize the difficulties which the designers of such structures had to overcome.

With the introduction of iron and steel for ship-building pur-

THE VICTORY

poses the necessity for ransacking the forests of the world for timber suitable for the frames and beam-knees of ships passed away, and Great Britain, which early became, and thus far remains, first and greatest in the production of iron and steel, was thus invited to such a development of naval power as the world has never seen. The mercantile marine of England at the present time furnishes a splendid demonstration of the readiness with which the commercial classes have appreciated this great opportunity; but the Royal Navy, by almost universal assent, supplies a melancholy counter-demonstration, and shows that neither the capabilities of a race nor the leadings of Providence suffice to keep a nation in its true position when it falls into the hands of feeble and visionary administrators. Anyone who will contrast the British navy of today, (as at time of first publication), with the British navy as it might and would have been under the administration, say, of such a First Lord of the Admiralty as the present Duke of Somerset proved himself in every department of the naval service five-and-twenty years ago, will understand the recent outcry in England for a safer and more powerful fleet.

It is impossible, as will presently appear, to describe the existing British navy without making reference to those administrative causes which have so largely and so unhappily influenced it; but the primary object of this chapter is, nevertheless, to describe and explain it, and only such references will be made to other circumstances as are indispensable to the fulfilment of that object.

It is fitting, and to the present writer it is agreeable, in this place, to take early note of a matter which has, perhaps, never before been fully acknowledged, *viz.*, the indebtedness of Great Britain and of Europe to the United States for some invaluable lessons in naval construction and naval warfare which were derived from the heroic efforts of their great civil war. The writer is in a position to speak with full knowledge on this point, as his service at the Admiralty, in charge of its naval construction, commenced during the American conflict, and continued for some years after its fortunate conclusion. There can be no doubt whatever that from the *Monitor* and her successors European constructors and naval officers derived some extremely valuable suggestions. The *Monitor* system itself, pure and simple, was never viewed with favour, and could never be adopted by England, except under the severest restrictions, because the work of England has mainly to be done upon the high seas and in distant parts of the world, and the extremely small freeboard of the *Monitor*, or, in other words, the normal submersion of so very much of the entire ship, is highly inconvenient and not a little dangerous on sea service, as the fate of the *Monitor* itself demonstrated. But for the work the *Monitor* was designed to do in inland waters she was admirably conceived, and her appearance in the field of naval warfare startled seamen and naval constructors everywhere, and gave their thoughts a wholly novel direction. In saying this I am not unmindful that seven years previously England had constructed steam-propelled "floating batteries," as they were called, sheathed with iron, and sent them to operate against the defences of Russia. But useful as these vessels were in many respects, their construction presented no striking novelty of design, and their employment was unattended by any dramatic incidents to powerfully impress the naval mind. The *Monitor* was both more novel and more fortunate, and opened her career (after a severe struggle for life at sea) with so notable a display of her offensive and defensive qualities that all eyes turned to the scene of her exploits, and scanned her with a degree of interest unknown to the then existing generation of sailors and ship-builders. Her form and character were in most respects singular, her low deck and erect revolving tower being altogether unexampled in steamship construction. He must have been a dull and conservative naval architect, indeed, whose thoughts Ericsson's wonderful little fighting ship did not stimulate into unwonted activity. But the serv-

The Glatton

ice rendered to Europe was not con-fined to the construction and exploits of the *Monitor* itself. The coasting passages, and, later on, the sea-voyages, of other vessels of the *Monitor* type, but of larger size, were watched with intense interest, and gave to the naval world instructive experiences which could in no other way have been acquired. Some of these experiences were purchased at the cost of the lives of gallant men, and that fact enhanced their value.

It is not possible to dwell at length upon the means by which the *Monitor* influence took effect in the navies of Europe, but it may be doubted whether ships like the *Thunderer*, *Devastation* and *Dreadnought*, which naval officers declare to be today the most formidable of all British warships, would have found their way so readily into existence if the *Monitors* of America had not encouraged such large departures from Old-world ideas. In this sense the *Times* correctly stated some years ago that the "American Monitors were certainly the progenitors of our *Devastation* type." The one ship in the British navy which comes nearest to the American Monitor, in respect of the nearness of her deck to the water, is the *Glatton*, a very exceptional vessel, and designed under a very peculiar stress of circumstances. But even in her case, as in that of every other armoured turret-ship of the present writer's design, the base of the turret and the hatchways over the machinery and boilers were protected by an armoured breastwork standing high above this low deck, whereas in the American Monitors the turret rests upon the deck, which is near to the smooth sea's surface.

We have here, in the features just contrasted, the expression of a fundamental difference of view between the American system, as applied to sea-going turret-ships, and the European system of sea-going ships introduced by the writer. It has never been possible, in our judgement on the British side of the Atlantic, to regard even such Monitors as the *Puritan* and *Dictator* were designed to be, as sufficiently proof to sea perils. At the time when these lines were penned the following paragraph appeared in English newspapers:

> The Cunard steamer *Servia* arrived at New York yesterday, being three days overdue. During a heavy sea the boats, the bridge, and the funnel were carried away, and the saloon was flooded.

Anyone who has seen the *Servia*, and observed the great height above the smooth sea's surface at which her boats, bridge, and funnel are carried, will be at no loss to infer why it is that we object to ships with upper decks within two or three feet only of that surface. In short, it can be demonstrated that ships of the latter type are liable, in certain possible seas, to be completely engulfed even to the very tops of their funnels. In the case of the *Glatton*, which had to be produced in conformity to ideas some of which were not those of the designer, one or two devices were resorted to expressly in order to secure in an indirect manner some increase of the assigned buoyancy, and thus to raise the upper deck above its pre-scribed height. The officers who served in her, however, judiciously regarded her, on account of Her low deck, as fit only for harbour service or restricted coast defence.

A very dangerous combination, as the writer regards it, was once proposed for his adoption by the representative of a colonial government, but was successfully resisted. This was the association of a "Coles" or English turret (which penetrates and passes bodily through the weather deck) with a low American Monitor deck. This was opposed on the ground that with such an arrangement there must of necessity be great danger at sea of serious leakage around the base of the turret as the waves swept over the lower deck. It would be extremely difficult to give to the long, circular aperture around the turret any protection which would be certain, while allowing the turret to revolve freely, both to withstand the tire of the guns and to resist the attack of the sea.

It will now be understood that while the *Monitor* system was from the first highly appreciated in Europe, and more especially in England, it never was adopted in its American form in the British navy. Russia, Holland, and some other powers did adopt it, and the Dutch government had to pay the penalty in the total disappearance of a ship and crew during a short passage in the North Sea from one home port to another. In a largely altered form, and with many modifications and additions due to English ideas of sea service, it was, however, substantially adopted in the three powerful ships already named, of which one, the *Dreadnought*, lately bore the flag of the British admiral who commands the Mediterranean fleet. If the opinion of officers who have served in these ships may be accepted as sufficiently conclusive, it was a great misfortune for the British navy when the ruling features of this type of ship were largely departed from in its first-class ships, and made to give place to a whole series of so-called first-class ironclads, of which only about one-third of the length has been protected by armour, and which are consequently quite unfit to take a place in any European line of battle.

The characteristic differences between the American type and the English type of sea-going Monitors (if we may apply that desig-

THE DREADNOUGHT

nation to the *Devastation* type) have already been stated, but may be restated here in a single sentence, *viz.*, the elevation in the English ship of the turret breastwork deck to a height of eleven or twelve feet above the sea's surface, and the raising of the upper deck generally, or of a considerable part of it, to at least that height, by means of lightly built superstructures. Over these again, and many feet above them, are built bridges and hurricane decks, from which the ships may be commanded in all weathers. Lofty as these ships are by comparison with American Monitors, it is only gradually that they have acquired the confidence of the naval service, so freely do the waves sweep over their weather decks when driven, even in moderate weather, against head-seas.

The British navy, having very diversified services to perform during both peace and war, requires ships of various kinds and sizes. Its first and greatest requirement of all is that of line-of-battle ships in sufficient numbers to enable England to stand up successfully against any European naval force or forces that may threaten her or her empire. If anyone should be disposed to ask why this requirement—which is obviously an extreme one, and an impossible one for more than a single power—is more necessary for England than for any other country, the answer must be, *Circumspice!* To look round over England's empire is to see why her failure on the sea would be her failure altogether. France, Germany, Italy, and even Holland, might each get along fairly well, losing nothing that is absolutely essential to their existence, even if every port belonging to them were sealed by an enemy's squadron. But were Great Britain to be cut off from her colonies and dependencies, were her ships to be swept from the seas, and her ports closed by hostile squadrons, she would either be deprived of the very elements of life itself, or would have to seek from the compassion of her foes the bare means of existence. It is this consideration, and the strong parental care which she feels for her colonies, that make her sons, indignant at any hazardous reduction of her naval strength. There are even in England itself men who cannot or will not see this danger, and who impute to those who strive to avert it ambitious, selfish, and even sordid motives. But it is to no unworthy cause that England's naval anxieties are due. We have no desire for war; we do not hunger for further naval fame; we cherish no mean rivalry of other powers who seek to colonize or to otherwise improve their trade; we do not want the mastery of the seas for any commercial objects that are exclusively our own. What we desire to do is to keep the seas open thoroughfares to our vast

The Inflexible

possessions and dependencies, and free to that commercial communication which has become indispensable to our existence as an empire. To accomplish that object we must, at any cost, be strong, supremely strong, in European waters; and it is for this reason that England's line-of-battle ships ought to be always above suspicion both in number and in quality.

It is not a pleasant assertion for an Englishman to make when he has to say that this is very far from being the case at present. A few months ago this statement, from whomsoever it emanated, would have been received with distrust by the general public, for the truth was only known to the navy itself and to comparatively few outsiders. But the official communications made to both Houses of parliament early in December, 1885, prepared the world for the truth, the First Lord of the Admiralty in the Chamber of Peers and Lord Brassey in the House of Commons having then proposed to parliament a programme of additional ship-building which provided for a considerable increase in the number of its first-class ships and cruisers, and which also provided, on the demand of the present writer, that the cruisers should be protected with belts of armour—an element of safety previously denied to them. It need hardly be repeated, after this wholesale admission of weakness by the Admiralty that Great Britain is at present in far from a satisfactory condition as regards both the number and the character of its ships. Were that not so, no public agitation could have moved the government to reverse in several respects a policy by which it had for so long abided.

It will be interesting to broadly but briefly review the causes of the present deplorable condition of the British navy. In the first place, in so far as it is a financial question, it has resulted mainly from the sustained attempt of successive governments to keep the naval expenditure within or near to a fixed annual amount, notwithstanding the palpable fact that every branch of the naval service, like most other services, is unavoidably increasing in cost, while the necessities of the empire are likewise unavoidably increasing. The consequence is that, as officers and men of every description must be paid, and all the charges connected therewith must in any event be fully met, the ship-building votes of various kinds are those upon which the main stress of financial pressure must fall. From this follows a strong desire, to which all Boards of Admiralty too readily yield, to keep down the size and cost of their first-class ships, to the sacrifice of their necessary qualities. This may be strikingly illustrated by the fact that, although the iron *Dreadnought*, a first-class ship, designed fifteen or sixteen years ago, had a displacement of 10,820 tons, and was powerful in proportion, the Admiralty has launched but a single ship (the *Inflexible*) since that period, of which the displacement has reached 10,000 tons. In fact, every large ironclad ship for the British navy since launched has fallen from twelve hundred to twenty-four hundred tons short of the *Dreadnought*'s displacement, and has been proportionally feeble.

If this cutting down in the size of the principal ships of Great Britain had been attended by a corresponding reduction in the sizes of the ships of other powers, or even by some advantages of design which largely tended to make up for the defect of size, there might be something to say for it. But the French ships have shown no such falling off in size, and have benefited as fully as the English ships by the use of steel and by the improved power and economy of the marine steam-engine.

Simultaneously with the reduction in the size of the English ships there has been brought about—voluntarily, and not as a consequence of reduced size, for it was first applied in the largest of all British men-of-war, the *Inflexible*—a system of stripping the so-called armoured ships of the English navy of a large part of their armour, and reducing its extent to so deplorable a degree that, as has already been said, they are quite unfit to take part, with any reasonable hope of success, in any general engagement. Here, again, there might have been something to say for a large reduction in the armoured surface of ships if it had been attended by some great compensation, such as that which an immense increase in the thickness of the armour applied might have provided, although no such increase could ever have compensated for such a reduction of the armoured part of the ship as would have exposed the whole ship to destruction by the mere bursting in of the unarmoured ends, which is what has been done. But although in the case of the large *Inflexible* the citadel armour was of excessive thickness, that is not true of the more recent ships of England, the armour of which sometimes falls short of that of the French ships, in two or three instances by as much as four inches, the French ships having 22-inch armour, and the English 18-inch. But by the combined effect of injudicious economy and of erroneous design, therefore—both furthered by a sort of frenzied desire on the part of the British Admiralty to strip the ships of armour, keep down their speed, delay their completion, and otherwise paralyze the naval service, apparently without understanding what they were about—the British navy has been brought into a condition which none but the possible enemies of the country can regard without more or less dismay.

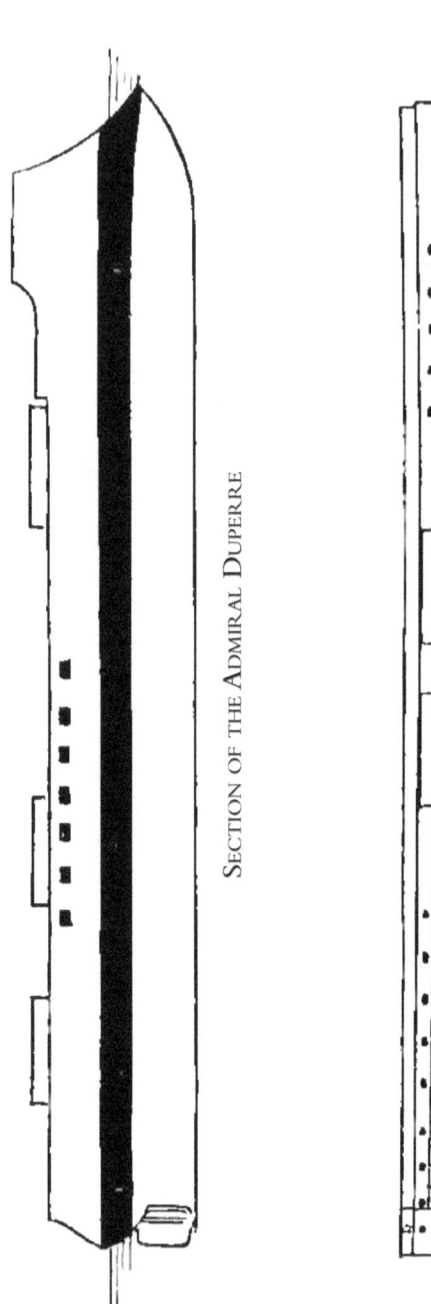

SECTION OF THE ADMIRAL DUPERRE

SECTION OF THE INFLEXIBLE

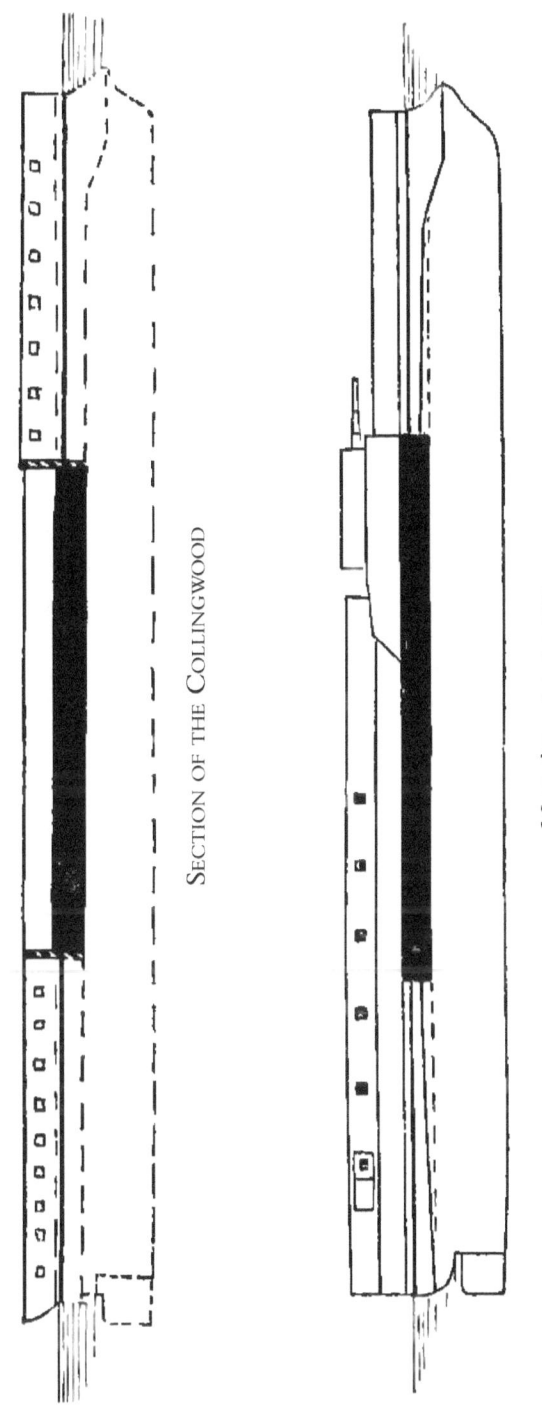

Section of the Collingwood

New Admiralty ship

In order to illustrate the extent to which side armour has been denied to the British ships, as compared with the French, we refer the reader to these diagrams of the *Amiral Duperré* (French) and of the *Inflexible* and *Collingwood* (both English). The black portions represent the side armour in each case. It is scarcely possible for any one friendly to Great Britain to look at these diagrams, and realize what they signify, without profoundly regretting that a sufficient force of public opinion has not yet been exerted to compel the Admiralty to a much more liberal use of armour in the new first-class ships, the intended construction of which was announced to parliament in December, 1885.

In these new ships, while the length of the partial belt has been slightly increased, no addition to its height above water has been made (as compared with the *Collingwood* or "Admiral" class), so that the slightest "list" towards either side puts all the armour below water. To describe such ships as "armoured ships" is to convey a totally false impression of their true character. A side view of one of these new ships shows that the two principal guns are carried high up forward in an armoured turret, which sweeps from right ahead, round the bow on each side, and well towards the stern, while several smaller guns are carried abaft with very thin armour protection to complete the offensive powers of the ship. The arrangement of the two principal guns in a turret forward resembles that of the *Conqueror*, but in her the armour rises high above the water, and a belt extends to the bow and nearly to the stern. It is a matter of inexpressible regret that the armoured surface of these new ships is so excessively contracted as to be wholly insufficient to preserve the ship from that terrible danger to which so many of their predecessors have been exposed, *viz.*, that of capsizing from loss of stability when the unarmoured parts alone have been injured.

There is a sense in which all the British ships to which reference has thus far been made may be roughly regarded as developments of, or at least as starting from, the *Devastation*, or British Monitor type of ship, for in all of them masts and sails have been done away with, and steam propulsion relied upon, a single military mast alone remaining. (This is not strictly true of quite all the ships named, but it probably will be true ere long, as none of them has more than a light auxiliary rig, and that will probably be abandoned.) We have now to notice another and more numerous class of ships, which may be regarded as the lingering representatives of those sailing-ships which have come down to us through the long centuries, but which are now rapidly disap-

THE DEVASTATION

pearing, yielding to the all-prevalent power of steam. Some of these ships were built for the line of battle, in their respective periods, but as they range in size from about one thousand tons of displacement up to nearly eleven thousand tons, it is obvious that many of them were built for various other employments. In dealing with the full-rigged ships, we are taking account of types of warships which, for all but secondary purposes, are passing away. It fell to the lot of the present writer (under the rule of Mr. Childers, then First Lord of the Admiralty, and of Admiral Sir Robert Spencer Robinson, then Controller of the Navy) to introduce the mastless warship, and thus to virtually terminate what had certainly been for England a glorious period, *viz.*, that of the taunt-masted, full-rigged, and ever-beautiful wooden line-of-battle ship. It is now, alas! but too apparent (from what has gone before) that in virtually terminating that period, and opening the era of the steam and steel fighting engine, we were also introducing an era in which fantastic and feeble people might but too easily convert what ought to have been the latest and greatest glory of England into her direct peril, and possibly even her early overthrow.

The first British ironclad (neglecting the "floating batteries" of 1854) was the *Warrior*, a handsome ship 380 feet long, furnished with steam-power, and provided with masts, spars, and a large spread of canvas. Her ends were unprotected by armour, and her steering gear consequently much exposed. She was succeeded by a long series of full-rigged ironclads, all of them supplied with steam-power likewise, the series continuing down to the present time. The little dependence which is now placed in the British navy upon the use of sail-power in armoured ships will be seen, however, when it is stated that of all the ships protected by side armour which are now under construction in the royal dockyards, but two are to be given any sail-power at all, and these are to be rigged on two masts only, although the ships are of large size, and intended for cruising in distant seas.

It is unnecessary in a popular subject of this description to dwell upon, or even to state, the minor differences which exist between the different types of rigged ironclads. There are, however, some points of interest in connection with their armour and armament to be mentioned. In the design of the first group (speaking chronologically) were commenced those changes in the disposition of the armour which continue down to the present time, the British Admiralty being so mixed and so virtually irresponsible a body that it is not obliged to have a mind of its own for any great length of time, even when many of the same men continue in office.

The *Warrior*, as we saw, and the sister ship *Black Prince*, had a central armoured battery only; the same is true of those reduced *Warriors*, the *Defence* and the *Resistance*. But the next succeeding ships of the *Warrior*'s size, the *Minotaur* and *Agincourt*, were fully armoured from end to end; and the somewhat smaller ship the *Achilles* was furnished with a complete belt at the water-line. The *Hector* and *Valiant* (improved *Defences*) had complete armour above the water, but, oddly enough, had part of the water-line at each end left unarmoured. A third ship of the *Minotaur* class, the *Northumberland*, was modified by the present writer at the bow and stern on his entering the Admiralty, the armour above water being there reduced, and an armoured bow breastwork constructed. Within this armoured breastwork were placed two heavy guns firing right ahead. With this exception, all these early ships, nine in number, were without any other protected guns than those of the broadside.

These ships were followed by a series of rigged ships of the writer's design, *viz.*, the *Bellerophon, Hercules, Sultan, Penelope, Invincible, Iron Duke, Vanguard, Swiftsure,* and *Triumph*, all with hulls of iron, or of iron and steel combined, together with a series of rigged ships constructed of wood, converted from unarmoured hulls or frames, *viz., Enterprise, Research, Favorite, Pallas, Lord Warden, Lord Clyde,* and *Repulse*. Every one of these ships was protected by armour throughout the entire length of the vessel in the region of the water-line, and in some cases the armour rose up to the upper deck. Most of them, however, had the armour above the belt limited to a central battery. The chief interest in these vessels now lies in the illustrations they furnish of the evolution, so to speak, of bow and stern fire. In several of them a fire approximately ahead and astern (reaching to those directions within about twenty degrees) was obtained by means of ports cut near to the ship's side, through the transverse armoured bulkheads. In others these bulkheads were turned inward towards the battery near the sides of the ship in order to facilitate the working of the guns when firing as nearly ahead and astern as was practicable. In the *Sultan* an upper-deck armoured battery was adopted for the double purpose of forming a redoubt from which the ship could be manoeuvred and fought in action, and of providing a direct stern fire from protected guns. In the five ships of the *Invincible* class a direct head and stern fire was obtained from a somewhat similar upper-deck battery, which projected a few feet beyond the side of the ship.

The rigged ships of later design than the writer's present a still greater variety in the disposition of their armour and armaments. This

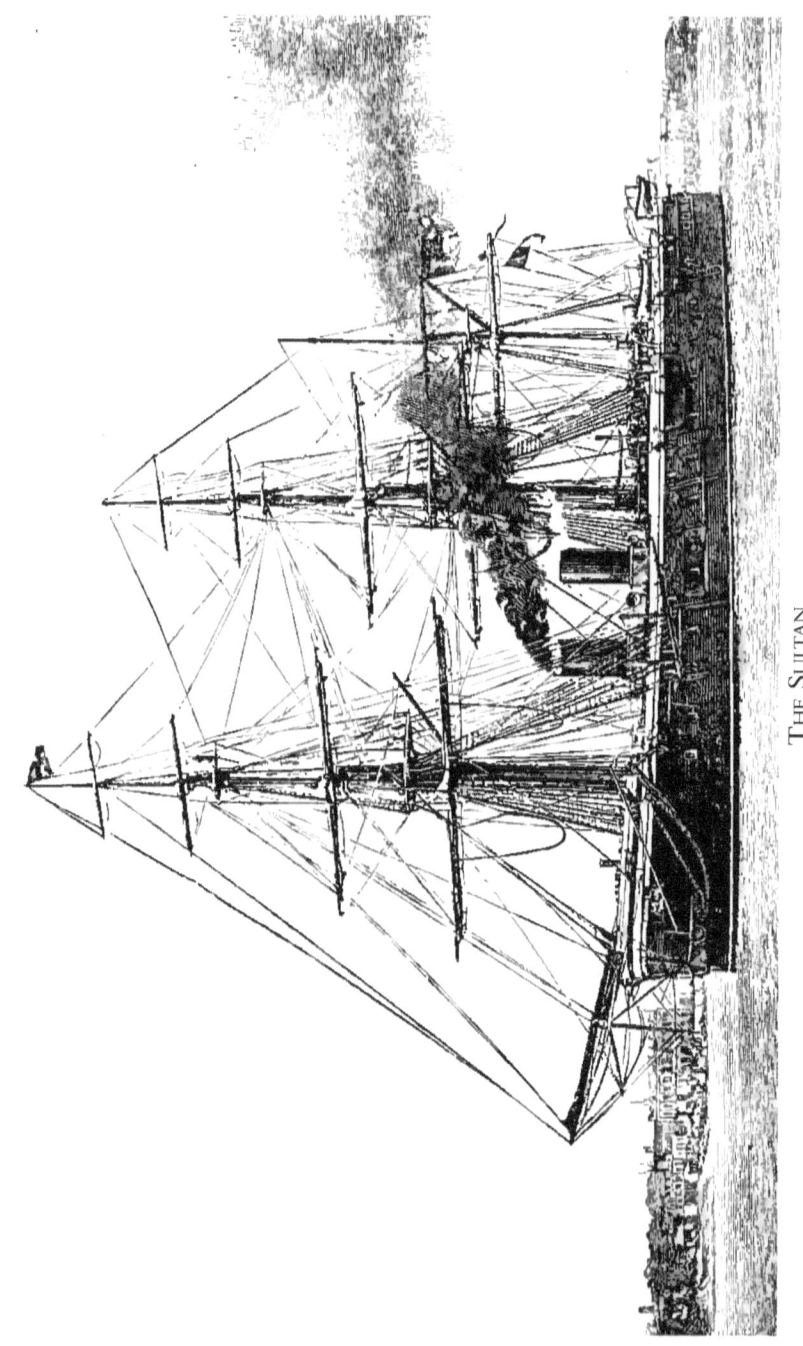

THE SULTAN

variety may be in part illustrated by four examples, which for convenience are principally taken from Lord Brassey's book *The British Navy*. The scales of these small drawings, as given there, are not all the same. These examples are the *Alexandra*, the *Téméraire*, the *Nelson*, and the *Shannon*. The *Alexandra* (of which a separate view, in sea-going condition, is given), which is probably the best of the rigged ironclads of the British navy, may be regarded as a natural, but not the less meritorious, development of the combined broadside and bow and stern fire of the central battery ships which preceded her. In her were provided a broadside battery on the main-deck, a direct bow fire, also on that deck, and both a direct bow and a direct stern fire on the upper deck from within armour, as in the *Invincible* class. The guns employed for bow and stern fire were all available for broadside fire. The upper-deck battery did not project beyond the main-deck as in the *Invincible* class, the forward and after parts of the ship above the main-deck being greatly contracted in breadth in order to allow the guns to fire clear both forward and aft. The *Téméraire* is a smaller ship than the *Alexandra*, and has a battery similar to hers on the main-deck, but with one gun less on each side, the danger of a raking fire entering through the foremost battery port being met by a transverse armoured bulkhead, as shown in the plan of the ship. She is provided with an additional bow gun and a stern-chaser, carried high up in barbette towers, but worked on Colonel Moncrieff's disappearing principle.

> The *Téméraire* fires three 25-ton guns right ahead, against two 25-ton and two 18-ton guns in the *Alexandra*; on either bow, two 25-ton against one 25-ton and one 18-ton; right aft, one 25-ton against two 18-ton; on either quarter, one 25-ton against one 18-ton; on either beam, if engaged on one side at a time, two 25-ton and two 18-ton, with a third 25-ton available through only half the usual arc, against three 18-ton guns, with two of the same weight and one of 25-tons, each available with the limitation just described. (From *Engineering*)

The *Alexandra* is a ship of 9500 tons displacement, the *Téméraire* is of 8500 tons; after them came the *Nelson* (to which the *Northampton* is a sister ship), of 7320 tons displacement. This vessel cannot be regarded as an armoured ship at all, in the usual sense of the word, having but a partial belt of armour, and none of her guns being enclosed within armour protection, although two guns for firing ahead and two for firing astern are partially sheltered by armour. Even less protection than this is afforded to the guns of the *Shannon*, which also

SECTION AND PLAN OF THE ALEXANDRA

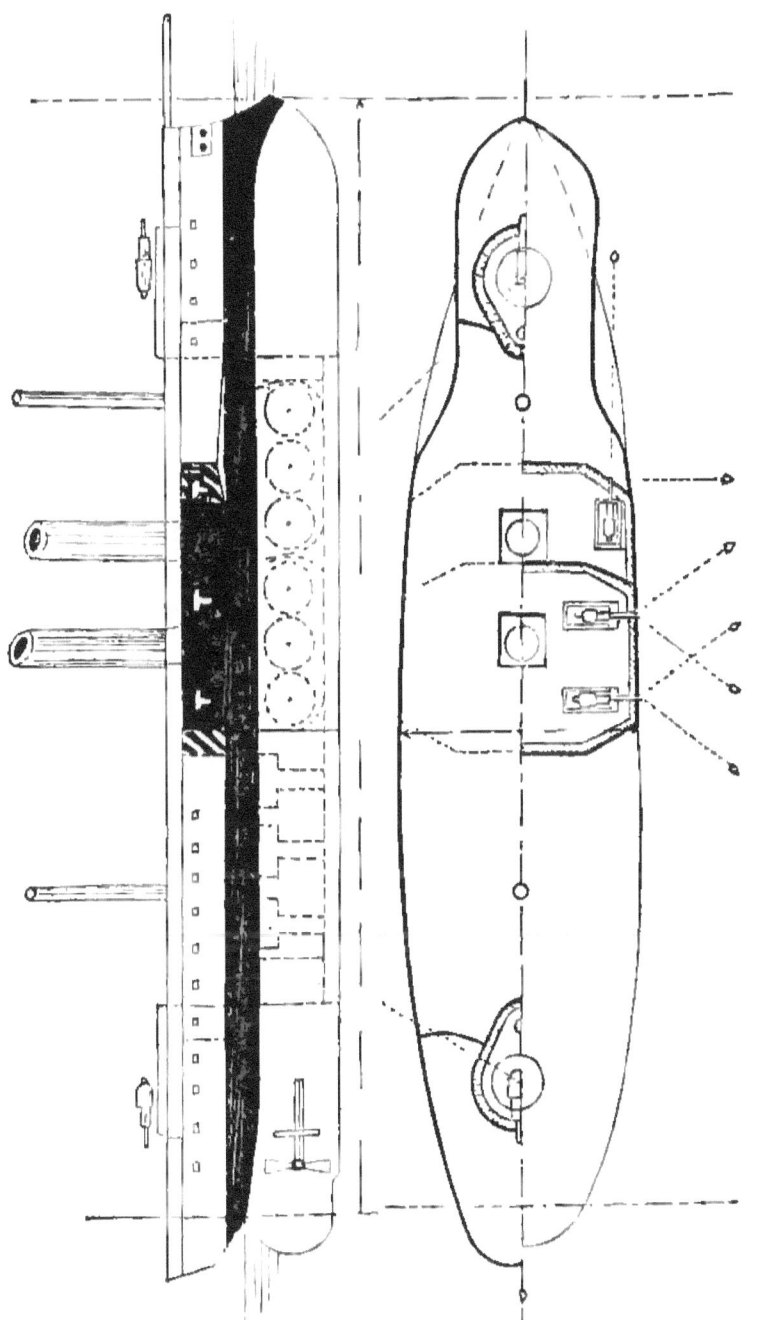

Section and plan of the Temeraire

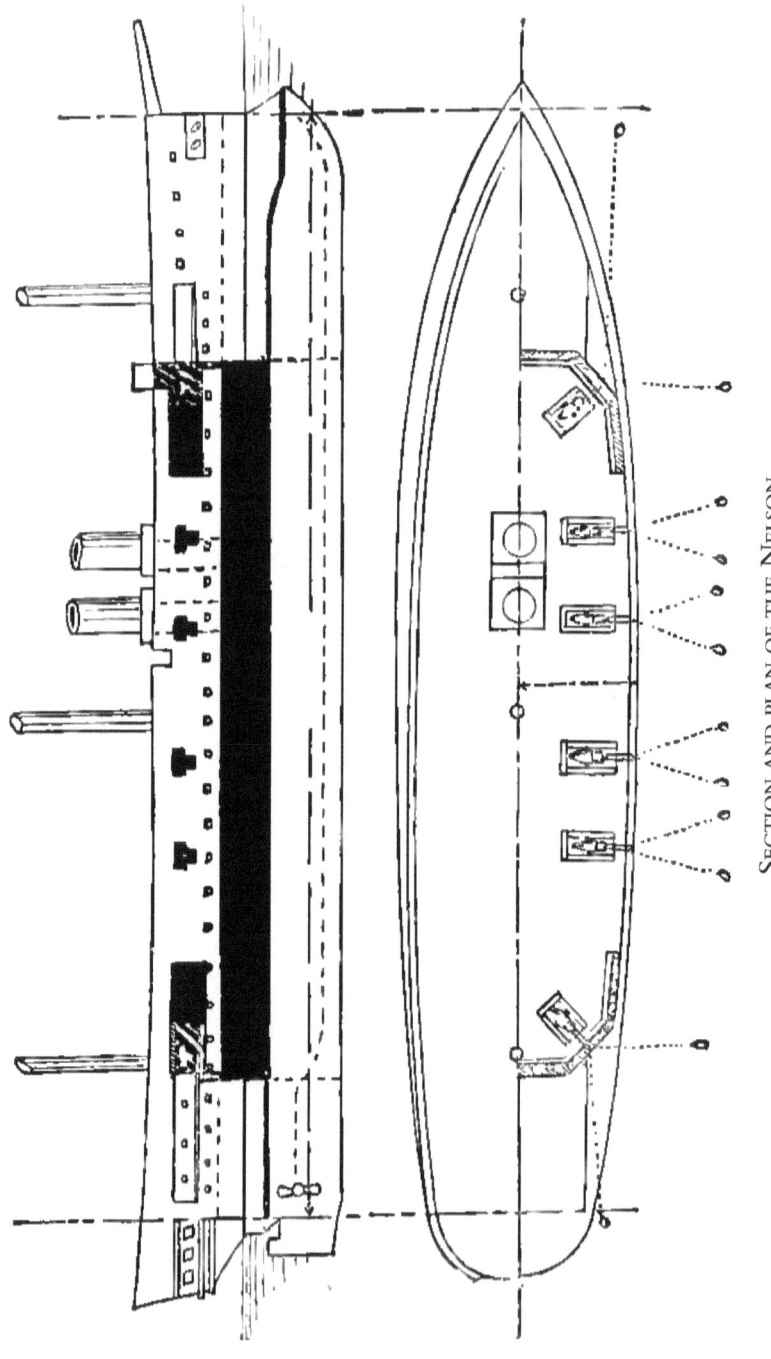

Section and Plan of the Nelson

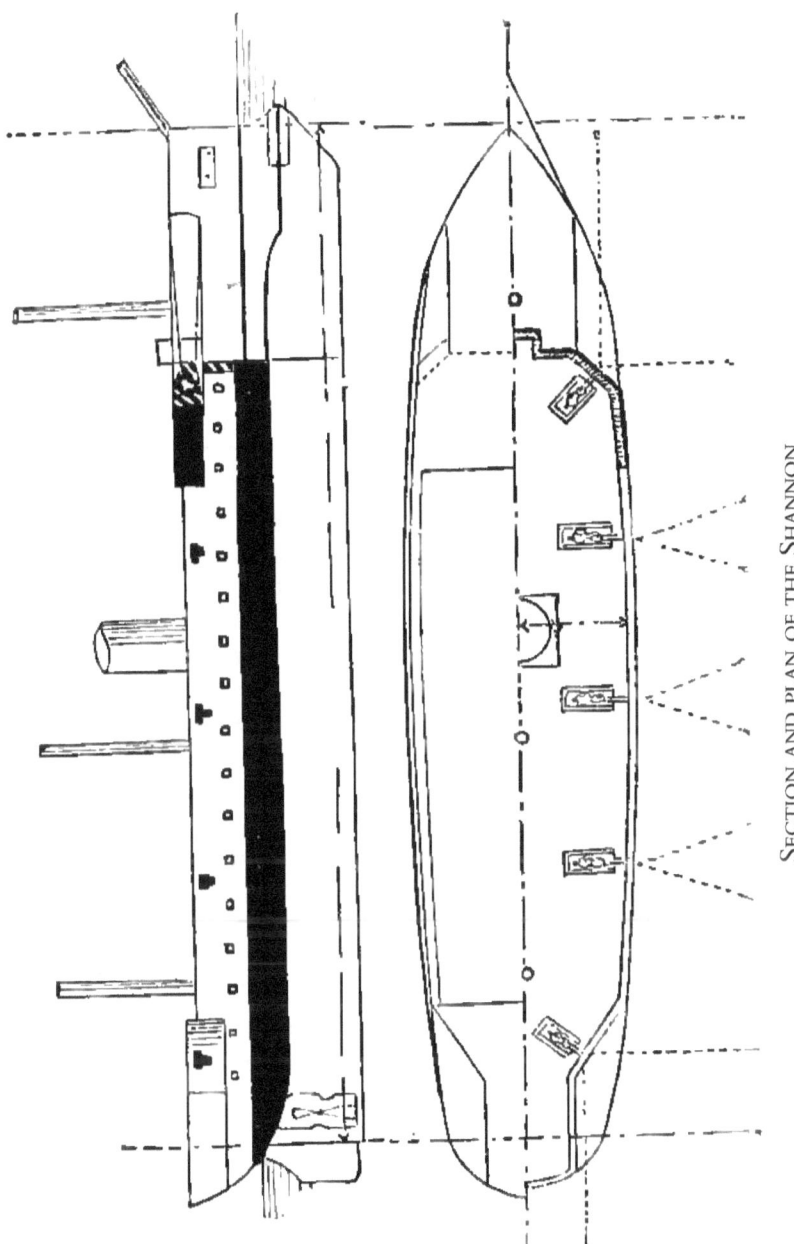

SECTION AND PLAN OF THE SHANNON

has but a partial belt of armour, and protection for two bow guns only. The comparatively small size of the *Shannon* (5400 tons displacement) relieves her in some degree from the reproach of being so little protected; but it is difficult (to the present writer) to find a justification for building ships of 7320 tons, like the *Nelson* and *Northampton*, and placing them in the category of armour-plated ships, seeing that their entire batteries are open to the free entrance of shell fire from all guns, small as well as large. "Where a ship has a battery of guns protected against fire in one or more directions, but freely exposed to fire coming in other directions, to assume that the enemy will be most likely to attack the armour, and avoid firing into the open battery, appears to be a reversal of the safe and well-accepted principle of warfare, *viz.*, that your enemy will at least endeavour to attack your vulnerable part. No doubt, when the size or cost of a particular ship is limited, the designer has to make a choice of evils, but where people are as free as is the British Board of Admiralty to build safe and efficient ships, the devotion of so much armour as the *Nelson* and *Northampton* carry to so limited a measure of protection is a very singular proceeding, and illustrates once more with how little wisdom the world is governed.

Before passing from the armoured ships of the navy—or, rather, as we must now say, in view of some of the ships just described and illustrated, before passing from the ships which have some armour—it is desirable to take note of a few exceptional vessels which cannot be classed either with the pretentious and so-called line-of-battle ships or with the rigged ironclads generally. Among these will be found two comparatively small ships, designed by the writer many years ago to serve primarily as rams, but to carry also some guns. These were the *Hotspur* and *Rupert*. The water-line of the *Hotspur* was protected with very thick armour for her day (11-inch), extending from stem to stern, dipping down forward to greatly strengthen the projecting ram. She carried (besides a few smaller guns) the largest gun of the period, one of twenty-five tons, mounted on a turntable, but protected by a fixed tower pierced with four ports. Some persons regarded the existence of these four small port-holes as converting the tower into a nest for projectiles, although a single enemy could not possibly have attacked more than two of these ports at once, situated as they were. What would such persons think of the batteries of the *Nelson*, *Northampton*, and *Shannon*, each open for more than one hundred feet in length, on each side of the ship, in so far as armour is concerned? This fixed tower was years afterwards replaced by a revolving turret, similar to that which the writer gave in the first instance to the *Rupert*, de-

The Alexandra

signed soon after the *Hotspur*. Both the armour and the armament of the second vessel were heavier than those of the first, but the ram, as before, was the chief feature of the ship.

It is needless here to describe some of the very early turret-ships, such as the *Prince Albert, Scorpion, Wyvern,* and *Royal Sovereign,* all of which embodied the early (though not by any means the earliest) views of that able, energetic, and lamented officer, the late Captain Cowper Coles, R.N., who was lost at sea by the capsizing of his own ship, the *Captain,* her low sides failing to furnish the necessary stability for enabling her to resist, when under her canvas, the force of a moderate gale of wind. Had he been able to foresee the coming abandonment of sail-power in rigged ships, and had he been placed, as the writer advised, in charge of the revolving turrets of the navy, leaving ship-designing to those who understood it, he might have been alive to this day, to witness the very general adoption in the British navy of that turret system to which he for some years devoted and eventually sacrificed his life.

The first real sea-going and successful ship designed and built to carry the revolving turret of Coles was, by universal consent, the *Monarch,* whose sea-going qualities secured for her the distinction of transporting to the shores of America—as a mark of England's goodwill to the people of the United States, and of her admiration of a great and good citizen—the body of the late Mr, George Peabody. Brassey's *British Navy* says:

> The performances of the *Monarch* at sea, were in the highest degree satisfactory.

Nothing could exceed the frank and liberal praises bestowed upon her for her performances during the voyage to New York by the officers of the United States man-of-war which accompanied her as a complimentary escort.

A great deal has been written and said at different times about four other turret-ships of the British navy, *viz.,* the *Cyclops, Gorgon, Hecate,* and *Hydra*—far less terrible vessels than these formidable names would seem to import. Whether these four comparatively small turret-ships possess the necessary sea-going qualities for coast defence (as distinguished from harbour service) is a question which has been much discussed, and is not yet settled. The truth is that the defence of the coasts of England, Wales, Scotland, and Ireland is a service in which the sea-going qualities of vessels may be called into requisition as largely as in any service in the world. There are some (this writer among them)

The Téméraire

who much prefer the mid-Atlantic in a heavy gale of wind to many parts of these coasts, more especially if there be any doubt about the perfect obedience of the ship to her steam-power and her helm. The worst weather the writer has ever experienced at sea was met with in the English Channel, and the only merchant-ship which he ever even in part possessed was mastered by a Channel storm, had to cast anchor outside of Plymouth Breakwater, was blown clean over it, and sank inside of it, with her cables stretched across that fine engineering work. It is therefore difficult, and has always been difficult, not to say impossible, for him to regard a "coast-defence ship," which certainly ought to be able to defend the coast, and to proceed from one part of it to another, as a vessel which may be made less sea-worthy than other vessels. Only in one respect, *viz.*, that of coal supply, may such a ship be safely made inferior to sea-going ships.

But whether the four vessels under notice be fit for coast defence or not, it ought to be known that they were not designed for it. They were hastily ordered in 1870, when the Franco-German war was breaking out, under the impression that Great Britain might get involved in that war. The British Admiralty knew then (as it knows now, and as it has known for years past) that the navy had not been maintained in sufficient strength, and it consequently seized the first design for a small and cheap ship that it could lay hands on, and ordered the construction, with all despatch, of four such vessels. The design which it happened to take, or which seemed to it most suitable, was that of the *Cerberus*—a breastwork Monitor designed by the writer for special service in inland colonial waters, and made as powerful as was then possible on 3300 tons of displacement, both offensively and defensively, but with no necessity for, and no pretensions whatever to, sea-going qualities. It is scarcely to be supposed that four vessels having such an origin could be expected to take their place as sea-going ships of the British navy; nor could they, either, for reasons already suggested, be expected to possess any high qualities as vessels for the defence of

That land 'round whose resounding coasts
The rough sea circles.

The Admiralty which ordered their construction may possibly be able to state why it built them, but even that is not at all. certain. One of the evil results of mean economies in national enterprises in ordinary times is extravagant and aimless expenditure in times of necessity.

A later example of this kind of expenditure under very similar circumstances was furnished during Lord Beaconsfield's administration,

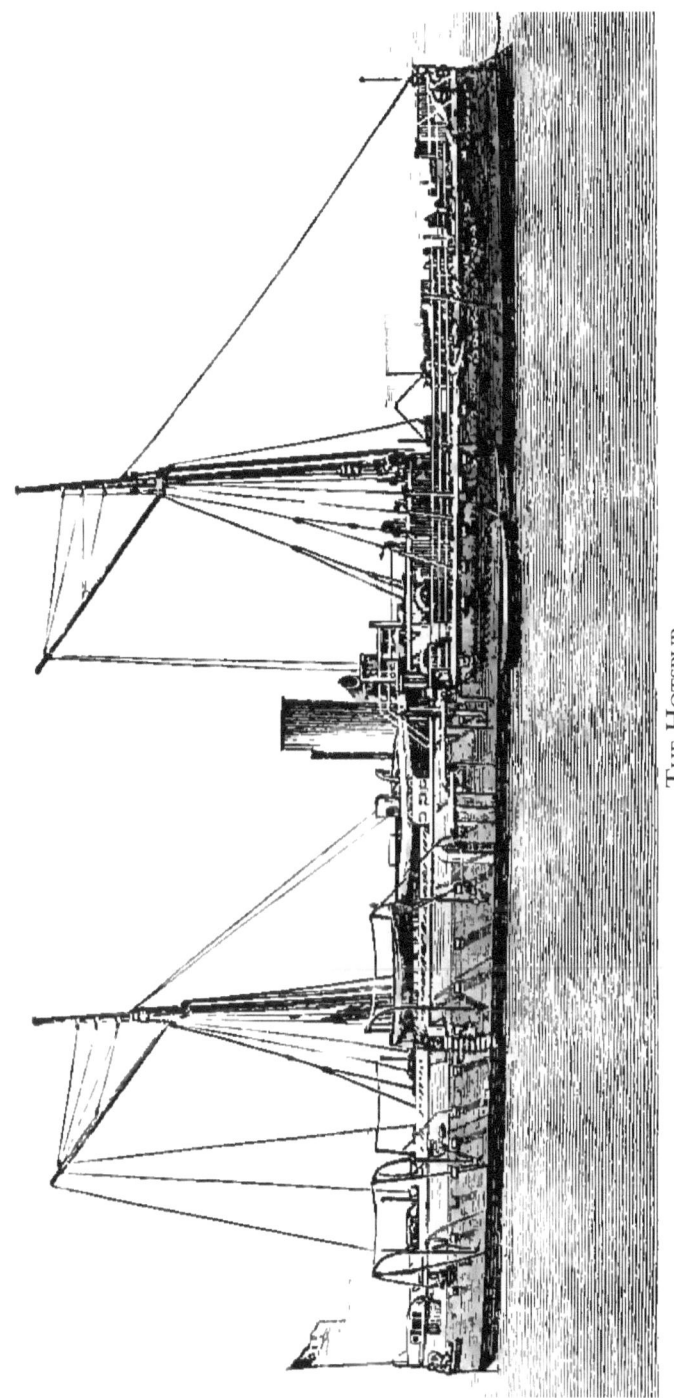

THE HOTSPUR.

when war with Russia seemed likely to occur. Again the insufficiency of the navy was strongly felt, and again public money to the extent of two millions sterling or more was expended upon the acquisition of such ships as could be most readily acquired, regardless of cost. At this time the *Neptune* (of 9170 tons displacement), the *Superb* (of 9,100 tons), and the *Belleisle* and *Orion* (each of 4830 tons), were purchased into the service, and having been built for other navies, and under very peculiar circumstances in some cases, required large dockyard expenditure to convert them to their new uses in the British navy.

It only remains, in so far as existing armoured, or rather "partly armoured," ships are concerned, to advert to the *Impérieuse* and *Warspite*, two cruisers building for distant service. These ships are three hundred and fifteen feet long, and to them has been allowed, by the extraordinary generosity of the Admiralty, as much as one hundred and forty feet of length of armoured belt. If this had been extended by only twenty feet, these British cruisers, which Lord Brassey—whether grandiloquently or satirically it is hard to say—calls "armoured cruisers," would have actually had one-half of their length protected by armour-plating at the water-line. In what spirit and with what object is not known, but Lord Brassey, in his outline sketch of these ships, writes the word "coals" in conspicuous letters before and abaft the belt. Can it be possible that he, undoubtedly a sensible man of business, and one who laboriously endeavours to bring up the knowledge and sense of his fellow-countrymen to a level with his own, and who was once Secretary to the British Admiralty—can it be possible that he considers coal a trustworthy substitute for armour, either before or after it has been consumed as fuel?

It is very distressing to have to write in these terms, and put these questions about Admiralty representatives and Admiralty ships; but what is to be done? Here are two ships which are together to cost nearly half a million of money, which are expressly built to chase and capture our enemies in distant seas, which are vauntingly described as "armoured cruisers," which cannot be expected always by their mere appearance to frighten the enemy into submission, like painted Chinese forts, which must be presumed sometimes to encounter a fighting foe, or at least to be fired at a few times by the stern guns of a vessel that is running away, and yet some eighty or ninety feet of the bows of these ships, and as much of their sterns, are deliberately deprived of the protection of armour, so that any shell from any gun may pierce them, let in the sea, and reduce their speed indefinitely; and in apparent justification of this perfectly ridiculous arrangement—perfectly

The Warspite

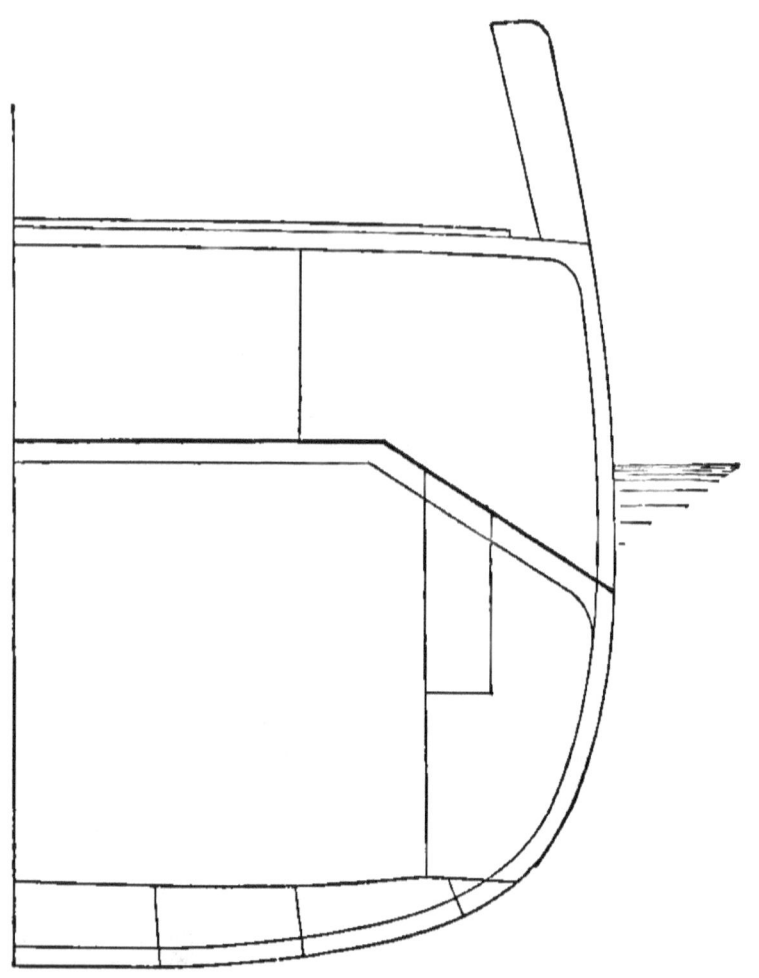

TRANSVERSE SECTION OF THE MERSEY

ridiculous in a ship which is primarily bound to sustain her speed when chasing—a late Secretary to the Admiralty tells us that she is to carry in the un-protected bow some coals! May my hope formerly expressed in *Harper's Magazine* find its fruition by giving to the British Admiralty a piece of information of which it only can be possibly ignorant, *viz.*, that even while coal is unconsumed, it differs largely from steel armour-plates in the measure of resistance which it offers to shot and shell; and further, that coal is put on board warships that it may be consumed in the generation of steam? It is very desirable that this information should somehow be conveyed to Whitehall in an impressive manner, and possibly, if the combined intelligence of the two great nations to which Harpers' publications chiefly appeal be invoked in its favour, it may at length be understood and attended to even by the Admiralty, and one may hear no more of the protection of her majesty's ships by means of their "coal."

Passing now from the so-called ironclads of the British navy, we come to a class of vessels which have their boilers, etc., protected from above by iron decks sweeping over them from side to side. The section of the *Mersey*, one of the most important British ships of this type, will illustrate the system of construction. Various attempts have been made to impose numerous ships of this kind upon a sometimes too credulous public as armoured vessels, and Lord Brassey, while publishing descriptions and drawings which demonstrated beyond all question that the buoyancy and stability of these ships are not at all protected by armour, nevertheless deliberately includes some of them, the *Italia* and *Lepanto*, for example, in his list of "armoured ships." Now, the thick iron deck certainly protects (in some degree, according to its thickness) all that is below it against the fire of guns, and armour itself is sometimes employed to protect the gun machinery; but the existence of a thickish deck under the water, or mainly under the water, occasionally associated with patches of armour above water here and there to protect individual parts, does not constitute the ship itself an armoured ship in any such sense of the term as is ordinarily accepted and understood. How can that be properly called an "armoured ship" which can be utterly destroyed by guns without any shot or shell ever touching such armour as it possesses?

The British Admiralty, in the *Navy Estimates* for 1883-84, under some unknown influence, put forward two ships of this description as armoured vessels, and was afterwards forced to remove them from that category, but only removed them to place them in another not less false, not less misleading, not less deceptive and dangerous, *viz.*, that of "protected ships." And this most improper description is still applied to various ships of which the special characteristic is that they themselves are *not* protected. If the ship's own coal and stores may be regarded as her protection, or if the existence of a certain number of exposed and extremely thin internal plates can be so regarded, then may these vessels be deemed partly, but only partly, "protected;" but if "protected ship" means, as every honest-minded person must take it to mean, that the ship herself is protected by armour against shot and shell, then the designation "protected ship," as employed by the British Admiralty, is nothing less than an imposition. These ships are not protected. Neither their power to float, nor their power to keep upright, nor their power to exist at all, after a few such injuries as even the smallest guns afloat can inflict, is "protected," as any war whatever is likely to demonstrate.

Those who employ such language ignore the essential characteristic of a ship-of-war, and some of the gravest dangers which menace her. It is conceivable that in the old days, when men wore armour, the protection of the head with an "armet," and of the breast by a breast-plate, might have justified the description of the man so defended as an "armoured man," although it is difficult to see why, since he might have been put *hors de combat* by a single stroke. But protect the boilers and magazines of a ship how you will, if you do not protect the ship itself sufficiently in the region of the water-line to prevent such an invasion of the sea as will sink or capsize her, she remains herself essentially unprotected, liable to speedy and complete destruction, and cannot truly be called a "protected ship."

It must not for a moment be supposed that this is a mere question of words or designations. On the contrary, it is one of the most vital importance to all navies, and most of all to the navy of Great Britain. What the Admiralty says, the rest of the government, and beyond them the country, are likely to believe and to rely upon, and when the stress of naval warfare comes, the nation which has confidently understood the Admiralty to mean "armoured ships" and "protected ships" when it has employed these phrases, and suddenly finds out, by defeat following defeat, and catastrophe catastrophe, that it meant nothing of the kind, may have to pay for its credulity, allowable and pardonable as it may be, the penalty of betrayal, and of something worse even than national humility.

On the other hand, it is not to be inferred from the objections thus offered to the employment of deceptive designations that objection is also offered to the construction of some ships with limited or partial protection, falling short of the protection of the buoyancy and the stability, and therefore of the life, of the ship itself. It is quite impossible that all the ships of a navy like that of Great Britain, or of the navies of many other powers, can be made invulnerable, even in the region of the water-line, to all shot or shell. Indeed, there are services upon which it is necessary to employ armed ships, but which do not demand the use of armoured or protected vessels. Unarmoured vessels, with some of their more vital contents protected, suffice for such services. Moreover, even where it would be very desirable indeed to have the hull protected by armour to a sufficient extent to preserve the ship's buoyancy and stability from ready destruction by gun-fire, it is often impracticable to give the ship that protection. This is true, for example, of all small corvettes, sloops, and gun-vessels, which are too small to float the necessary armour-plates, in addition to all the indispensable weights of hull, steam-machinery, fuel, armament, ammunition, crew, and stores. It would be both idle and unreasonable, therefore, to complain of the construction of some ships with the protecting armour limited, or even, in certain cases, with no protecting armour at all. Such ships must be built, and in considerable number, for the British navy. But this necessity should neither blind us to the exposure and destructibility of all such vessels, nor induce us to endeavour to keep that exposure and destructibility out of our own sight. Still less should it encourage us to sanction, even for a moment, such an abuse of terms as to hold up as "armoured" and "protected" ships those which, whether unavoidably or avoidably, have been deprived of the necessary amount of armour to keep them afloat under the fire of small or even of moderately powerful guns.

We are now in a position to review the British navy, and to see of what ships it really consists. In this review it will not be necessary to pass before the eyes of the reader that large number of vessels of which even the boilers and magazines are without any armour or thick-plate protection whatever. It will help, nevertheless, to make the nature and extent of the navy understood if these are grouped and summarized in a few sentences. Neglecting altogether all large vessels with timber frames (which may be regarded as out of date, seeing that all the war vessels of considerable size now built for the navy have iron or steel frames), it may be first said that there are but three ships of the large or frigate class in the British navy which carry no thick protecting plate at all, *viz.*, the *Inconstant*, the *Shah*, and the *Raleigh*. Of much less size than these, and equally devoid of protection, are the two very fast vessels, the *Iris* and *Mercury*, built as special despatch vessels, steaming at their best at about eighteen knots. Among the unarmoured corvettes are the *Active, Bacchante, Boadicea, Euryalus, Rover*, and *Volage*, all exceeding fourteen knots in speed, and all more than three thousand tons displacement. Then follow thirty-six smaller and less swift corvettes, nearly one-half the number being built wholly of wood, most of which exceed, however, thirteen knots in speed; and below these about an equal number of sloops of less speed and tonnage. The smaller gun-vessels and gun-boats need not be summarized.

Passing on to vessels which, although themselves unarmoured, have thick-plate decks to give some protection to the machinery, we observe first that there are eight ships of three thousand five hundred

The Inconstant

to three thousand seven hundred tons built and under construction, *viz.*, the *Amphion, Arethusa, Leander, Phaeton, Mersey, Severn, Forth*, and *Thames*. Lord Brassey very properly classes such of these vessels as he mentions in his lists as "unarmoured ships," although, as before mentioned, when two of them—the *Mersey* and *Severn*—were designed, with a deck two inches thick, the Admiralty at first ventured to put them forward as "armoured ships."

Ascending in the scale of protection, and dealing for the present with sea-going vessels only, we come to a long series of ships which are undeserving of the designation of armoured ships, because they are liable to destruction by guns without the limited amount of armour which they carry being attacked at all. These ships are the *Impérieuse* and *Warspite*, previously discussed, and also the *Ajax, Agamemnon, Colossus, Edinburgh*, and the six large ships of the "Admiral" class. Anyone who has intelligently perused the report of the committee on the *Inflexible* would justify the inclusion of that ship in this category; but she is omitted here out of deference to the strenuous exertions which were made to invent or devise some little stability for her, even when her bow and stern are supposed to be badly injured, and out of compassion upon those officers of the Admiralty who have long ago repented those trying compromises with conscience by aid of which they expressed some slight confidence in her ability to float upright with her unarmoured ends badly damaged. She is omitted also out of gratitude to Lord Brassey for a sentence in which, while saving her from being placed in so dreadful a category, he honestly places some of the other ships in it without qualification or circumlocution. He says:

> In one important particular the *Ajax* and *Agamemnon* are inferior to the *Inflexible*. The central armoured citadel is not, as it is in the case of the Inflexible, of sufficient displacement to secure the stability of the ship should the unarmoured ends be destroyed. (*The British Navy*, vol. i.)

In another place the former Secretary to the Admiralty, referring to the report of the *Inflexible* committee (which was nominated by the Admiralty, and under heavy obligations to support it), says:

> It is doubtless very desirable that our armoured ships should possess a more ample margin of stability than is provided in the armoured citadel of the *Inflexible*. The ideas of the committee and of Sir Edward Reed on this point were in entire accord.

It has recently been acknowledged that, as Lord Brassey states,

The Colossus

the *Ajax* and *Agamemnon* are so constructed that they are dependent for their ability to float, the right side uppermost, upon their unarmoured ends. To call such ships "armoured ships" is, as we have seen, to mislead the public. But some pains have been taken of late to show that the "Admiral" class is better off in this respect, and certainly the known opinions of the present writer have been so far respected in these ships that their armoured citadels, so called, have been made some-what longer and of greater proportionate area. The following figures have been given:

	Percentage of water-line area covered by armor.
Inflexible	42.
Agamemnon	45.4
Collingwood	54.15
Camperdown	56.35

But anyone who understands this question knows perfectly well that "percentage of water-line area covered by armour" in no way represents the relative stabilities of these ships. Indeed, that is obvious upon the face of the matter, because we have seen the *Ajax* and *Agamemnon* pronounced devoid of the necessary stability when injured, while the *Inflexible* is said to possess it, although the former vessel has 45½ *per cent*, of the water-line area covered, while the latter has but 42 *per cent*. But this is not the consideration which has led to the condemnation of the whole "Admiral" class of so-called ironclads as not possessing the essential characteristic of an armoured ship, *viz*., the power to float, and to float with needful buoyancy and stability, all the time the armour is unpierced. The ground of that condemnation is to be found in the introduction into the "Admirals" of a dangerous combination from which the *Inflexible* and *Agamemnon* and other like ships are exempt—the combination of long unarmoured ends comprising about forty-five *per cent*, of the water-line area with so shallow a belt of armour that, when the unarmoured ends are injured and filled by the sea (as they would be in action), there would remain so little armour left above water that a very slight inclination of the ship would put it all below water. In the *Agamemnon* class, small as the initial stability may be (and with the unarmoured ends torn open it would be nothing), the armour is carried up to a reasonable height above water. But in the "Admiral" class all the advantage arising from a slightly lengthened citadel is more than destroyed by this lowering of the armour. So great is the consequent danger of these ships capsizing, if ever called upon to engage in a serious battle at close quarters, that the writer cannot conscientiously regard them as "armoured ships," but must in common fairness to the officers and men who are to serve in them, and to the nation which might otherwise put its trust in them, relegate them to the category of ships with only parts protected.

It will be observed that nothing has yet been said about thickness of armour, although that is, of course, a very important element of a ship's safety or danger. But important as it is, it has to be kept scrupulously separated from the question just discussed—the limitation of the armour's extent—because no misrepresentation and no misconception can well arise concerning the relative power or trustworthiness of ships armoured variously as to thickness, while much misrepresentation has actually taken place, and much consequent misconception has actually arisen, on the other matter, more than one European government having deliberately placed in the category of "armoured ships" ships which in no true sense of the word can be so classed.

The following classifications will conform to the foregoing views, describing as "armoured ships" only those which have sufficient side-armour to protect them from being sunk or capsized by the fire of guns all the time the armour remains unpierced:

BRITISH SHIPS OF WAR, BUILT AND BUILDING.

ARMORED SHIPS WITH THICK ARMOR.

Name of Ship	Tons Displacement	Indicated Horsepower	Speed, in Knots	Maximum Thickness of Armor, in Inches.	Largest Guns, in Tons.
Alexandra	9,490	8,610	15	12	25
Belleisle	4,830	3,200	12¼	12	25
Conqueror	5,200	4,500	15	12	43
Devastation	9,330	6,650	13¾	12	35
Dreadnought	10,820	8,200	14¼	14	38
Hero	6,200	4,500	15	12	43
Inflexible	11,400	8,000	14	24	80
Neptune	9,170	9,000	14⅘	12	38
Orion	4,830	3,900	13	12	25
Rupert	5,440	4,630	13½	12	18
Superb	9,100	7,480	14	12	18
Thunderer	9,330	6,270	13¼	12	38
Glatton	4,910	2,870	12	12	25

ARMORED SHIPS WITH MEDIUM ARMOR.

Name of Ship	Tons Displacement	Indicated Horsepower	Speed, in Knots	Maximum Thickness of Armor, in Inches.	Largest Guns, in Tons.
Hercules	8,680	8,530	14⅞	9	18
Hotspur	4,010	3,500	12¾	11	25
Sultan	9,290	8,630	14	9	18
Téméraire	8,540	7,700	14½	11	25

ARMORED SHIPS WITH THIN ARMOR.*

Name of Ship.	Tons Displacement.	Indicated Horse-power.	Speed, in Knots.	Maximum Thickness of Armor, in Inches.	Largest Guns, in Tons.
Achilles	9,820	5,720	14½	4½	12
Agincourt	10,690	6,870	15	5½	12
Audacious	6,910	4,020	13	8	12
Bellerophon	7,550	6,520	14¼	6	12
Black Prince	9,210	5,770	13¾	4½	9
Gorgon	3,480	1,650	11	9	18
Hecate	3,480	1,750	11	9	18
Hector	6,710	3,260	12½	4½	9
Hydra	3,480	1,470	11¼	8	18
Invincible	6,010	4,830	14	8	12
Iron Duke	6,010	4,270	13¾	8	12
Minotaur	10,690	6,700	14½	5½	12
Monarch	8,320	7,840	15	7	25
Northumberland	10,580	6,560	14	5½	12
Penelope	4,470	4,700	12¾	6	9
Prince Albert	3,880	2,130	11¾	4½	12
Swiftsure	6,640	4,910	15¾	8	12
Triumph	6,640	4,890	14	8	12
Valiant	6,710	3,560	12¾	4½	9
Warrior	9,210	5,470	9¼	4½	9

The ships in the list below, although having some armour upon their sides, being liable to capsize at sea from injuries inflicted upon their unarmoured parts, cannot be classed with the armoured ships.

SHIPS ARMORED IN PLACES.

Name of Ship	Tons Displacement	Indicated Horse-power.	Speed, in Knots.	Maximum Thickness of Armor, in Inches.	Largest Guns, in Tons.
Ajax	8,490	6,000	13	18	38
Agamemnon	8,490	6,000	13	18	38
Anson	10,000	7,500	14	18	63
Benbow	10,000	7,500	14	18	110
Camperdown	10,000	7,500	14	18	63
Collingwood	9,150	7,000	14	18	43
Colossus	9,150	6,000	14	18	43
Edinburgh	9,150	6,000	14	18	43
Howe	9,600	7,500	16	18	63
Rodney	9,600	7,500	14	18	63
Impérieuse	7,390	8,000	16	10	18
Warspite	7,390	8,000	16	10	18

To the preceding list may now be added two ships of 10,400 tons displacement, with 18-inch armour, and live cruisers of 5000 tons displacement, with 10-inch armour, recently ordered by the Admiralty to be built by contract.

Name of Ship	Tons Displacement.	Indicated Horsepower.	Speed in Knots.	Thickness of Deck, in Inches	Largest Guns, in Inches.
Amphion	3,750	5,000	$16\frac{3}{4}$	$1\frac{1}{2}$	6
Arethusa	3,750	5,000	$16\frac{3}{4}$	$1\frac{1}{2}$	6
Leander	3,750	5,000	$16\frac{3}{4}$	$1\frac{1}{2}$	6
Phaeton	3,750	5,000	$16\frac{3}{4}$	$1\frac{1}{2}$	6
Mersey	3,550	6,000	17	2	6
Severn	3,550	6,000	17	2	6
Thames	3,550	6,000	17	2	6
Forth	3,550	6,000	17	2	6

The thicknesses of decks given are those of the horizontal, or nearly horizontal, parts of the deck. Where the decks slope down at the sides the thickness is sometimes increased a little, as will have been seen in the section of the *Mersey*.

Armoured ships with 12-inch armour and upward are called ships with thick armour; those with armour less than twelve inches but more than eight inches thick are designated as ships with medium armour; and those with 8-inch armour or less as ships with thin armour.

A number of vessels of the "Scout" class are now under construction for the Admiralty. There is a disposition in certain quarters to include these among the ships of the class recorded in the last table. A transverse section of one of these is given here, in which the so-called protective deck is but three-eighths of an inch in thickness, and can therefore be pierced by any gun afloat, from the largest down to the very smallest. It would be quite absurd to speak of this class of vessels as being in any way "protected" against gunfire.

The first-class ships, so called, and the armoured cruisers referred to in the former part of this chapter as having been promised to parliament by the Admiralty representatives, were ordered, and work upon them is well under way in the yards of those firms to whom their building has been entrusted. The former are two in number, and their principal dimensions and particulars are as follows: length, 340 feet; breadth, 70 feet; draught of water, 26 feet; displacement, 10,400 tons; indicated horse-power, 10,000; estimated speed, 16 knots; thickness of armour, 18 inches; largest guns, 110 tons. The armour-belt in these ships is a little more than 160 feet long, or about half their length, but rises to a height of only two feet six inches above the water. Before and abaft the belt under-water armoured decks extend to the stem, and stern respectively, as in the "Admiral" class. Besides the two 110-ton guns, which, as has been

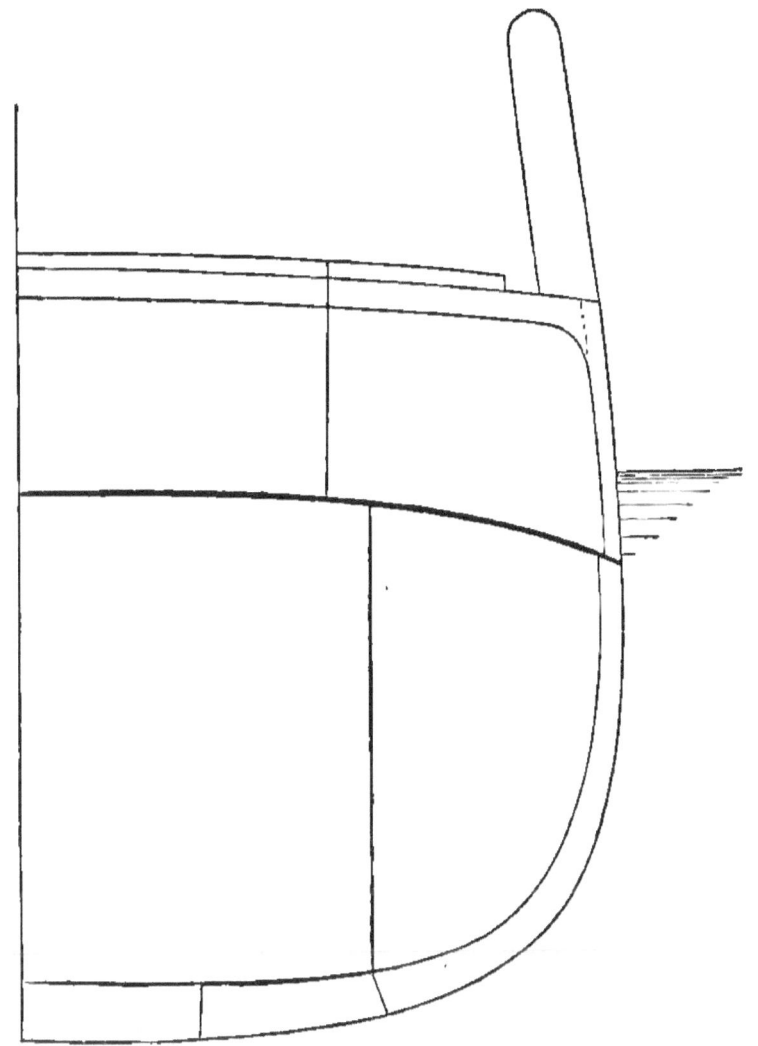

TRANSVERSE SECTION OF ONE OF THE NEW SCOUTS

said, are placed in a turret forward and fire over the upper deck, there are twelve 6-inch guns ranged round the after-part of the ship on the upper deck. A certain amount of protection has been given to these guns by means of armour-plating, but as this is only three inches thick, it can be said to do little more than protect the gun crews from the fire of rifles and of the smallest machine-guns.

Of the armoured cruisers, five have been contracted for. Their principal dimensions and particulars are: length, 300 feet; breadth, 56 feet; draught of water, 21 feet; displacement, 6000 tons 5 indicated horse-power, 8500; estimated speed, 18 knots; thickness of armour, 10 inches; largest guns, 18 tons. These vessels are protected by an armour-belt nearly two hundred feet long, which extends to a height of one foot six inches above the water, and to a depth of four feet below it, and they also have underwater decks before and abaft the belt. They carry two 18-ton guns, one well forward, ranging right round the bow, and the other well aft, ranging right round the stern, as well as five 6-inch guns on each broadside, the foremost and aftermost of which are placed on projecting sponsons, by which they are enabled to fire right ahead and right astern respectively. None of these guns is protected except by the thin shields usually fitted to keep off rifle fire from those actually working the guns.

No mention has yet been made of the troop or transport ships of the British navy. There are in all about a dozen of these, but by far the most conspicuous and important of them are the five Indian transports which were built about twenty years ago, conjointly by the Admiralty and the government of India, and ever since worked by those departments of the State with general satisfaction. One of these, the *Jumna*, is illustrated in the annexed figure. So satisfied was the late Director of Transports, Sir William R. Mends, K.C.B., with the services of these ships that, before retiring from his office, he informed the writer that if he had to assist in the construction of a new fleet of such transports he would desire but a single improvement in them, as working ships, and that was the raising of the lower deck one foot, in order to increase to that extent the stowage of the holds.

In the early part of this chapter the writer made reference to the influence exerted upon European ship-building by the incidents of the American civil war. He will conclude by a reference to an influence exerted upon his own mind and judgement by the most distinguished naval hero of that war, the late Admiral Farragut. On the occasion of that gallant officer's visit to England the Board of Admiralty invited him, as a wholly exceptional compliment, to accompany it on its annual official visit of inspection to her Majesty's dockyards. On the way from Chatham to Sheerness in the Admiralty yacht, the writer had a most instructive conversation with the admiral as to the results of his practical experience of naval warfare at the brilliant capture of New Orleans, and elsewhere, and one of those results was this:

> Never allow your men to be deceived as to the ships in which you expect them to fight. They will fight in anything, and fight

THE JUMNA

to the death, if they know beforehand what they are going about, and what is expected of them. But if you deceive them, and expose them to dangers of which they know nothing, and they find this out in battle, they are very apt to become bewildered, to lose heart all at once, and to fail you just when you most require their utmost exertions.

The writer has not forgotten this, and will not forget it. The British Admiralty is, unhappily, altogether unmindful of it.

Notes

There is no rigorous law by which a universal naval policy may be formulated, for a nation's environment, geographical and political, defines the conditions that must be obeyed. Underneath all, however, the immutable principle exists that the first and supreme duty of a navy is to protect its own coasts. The measures required to achieve this end are as various as a country's necessities, resources, opportunities, and temperament. England, for example, has always guarded her homes, not at the hearth-stone nor the threshold, but within gunshot of her enemy's territory; her defence has been an attack upon his inner line, and her vessels have been, not corsairs preying upon merchantmen, but battleships, ready for duel or for fleet engagement, whether they had the odds against them or not. This is the true sailor instinct; this has made England's greatness.

Today, (as at time of first publication), the question is so much governed by the complexities of modern progress that the details must be altered to suit the new demands; for it is not the England of the British Islands nor of the sparsely settled colonies that is now to be defended—it is a Greater Britain. The trade and commerce of England have increased so enormously in late years that no figures are necessary to show the interests she has afloat; but as proof of her growth in territory and in population outside of the mother-country, these statistics, taken from a late number of the *Nineteenth Century*, may perhaps be quoted:

FIFTY YEARS' GROWTH OF INDIA AND THE COLONIES

INDIA.

	1835.	1885.
Area governed in square miles	600,000	1,380,000
Population of European stock	300,000	500,000
Population, colored	96,000,000	254,000,000
State revenues	£19,000,000	£74,000,000

COLONIES AND DEPENDENCIES.

	1835.	1885.
Area governed in square miles	520,000	7,000,000
Population of European stock	1,800,000	9,500,000
Population, colored	2,100,000	8,000,000
State revenues	£5,000,000	£51,000,000

That is to say, in fifty years England has added 7,260,000 square miles to her territory, and nearly trebled the population she controls in India and her colonies. Is it necessary to add that with all this at stake the ocean highways which her ships traverse must be held toll free; that the nations which she has peopled and owns must be protected; that the enemy's squadrons which will seek to cut off her food supply, destroy her commerce, and burn her coaling stations, must be chased and captured; or that in the line of battle her ships must meet his and conquer? Sea-going and sea-keeping fleets and their auxiliaries must always be ready; transferable forts for protection abroad, and coast-defence ships for safety at home, must be kept afloat; and, in a word, every means must be employed which, through successful sea war, will maintain her integrity as a nation. Her navy must be eclectic in types, the exact instrument for any expected operation being always at hand; her maritime administration must be comprehensive; and her preparation ever such as will anticipate and surpass that of all her rivals. Enormously armoured battleships may be economically wrong, but while other countries build them so must she; for her immunity depends not upon treaties nor the friendly but false protestations of rivals, but upon the fear of her unassailable superiority, A mistaken naval policy is to any nation a grave disaster, but to England it means ruin. Lord Brassey: wrote:

> We cannot allow any foreign power to possess vessels which we cannot overhaul, or to carry guns at sea which may inflict a dam-aging blow to which it is impossible for us to reply. "We must have ships as fast as the fastest, and guns at least equal to the most powerful which are to be found in the hands of any possible enemy.

Knowing, then, the interests imperilled, English designers are keen to achieve the best results; and when, as they believe, this has been accomplished, is it a wonder that they fall tooth and nail in a white-heat of positive assertion and flat contradiction upon all who differ from them? All are striving so honestly for the common good of the great country which they love with such intense and insular patriotism that even their embittered differences of belief command the respect of right-thinking men everywhere. But in these variant faiths where is the truth? The question has run the gamut of experiment without being solved, the pendulum has swung from side to side and found no point of rest, and today there is a fixed agreement only as to the dangers which threaten.

The most marked tendencies, however, in all modern design are the diminution of side armour, the increase of deck protection, and the development of speed. The public mind is so familiar with the great speed of the large mail-boats that a common question, often put as an inquiry of disparaging comparison, is why war vessels do not steam as fast. The simplest answer to this is that they do, and in types which, like the big steamers, are special, the boasted achievements have been surpassed. Of course the number of vessels that can make nineteen knots is limited, because the man-of-war is hampered by necessities of space, weight, and safety, which do not obtain with the others. Mr. White, who has been so often, and, it is to be hoped, so advantageously, quoted in this editing, says:

> The necessity for giving protection to the engines and boilers of warships introduces special restrictions and difficulties in the design which are not known in merchant-ships, wherever in warships the overshadowing necessities of fighting power compel the acceptance in many cases of limited space and other inconveniences.... Merchant-steamers of all classes are built and engined for the purpose of steaming continuously at certain maximum speeds, and making fairly uniform passages; they consequently possess a considerable reserve of boiler power to meet adverse conditions of wind and sea. Warships, on the contrary, ordinarily cruise at very low speeds, and yet must be capable of reaching very high speeds when required in action or chasing. A warship, for instance, that attained about sixteen knots on the measured mile, and could steam continuously at sea, as long as her coal lasted, at a speed of about fifteen knots, would ordinarily have to cruise at from nine to ten knots. At this low speed she would require, say, only *one-seventh* of the indicated horse-power which would be developed at her full sea speed, or say *one-tenth* of what would be developed on the measured mile. This obviously introduces conditions of a character entirely different from those of the merchant-ship. The warship's machinery must be so designed that the power necessary to give her high speed at long

intervals and for short periods should be secured with the least expenditure of weight consistent with insuring the maximum performance when required, and with the provision of proper strength and durability.

The very vague ideas existing as to the cost of increased speed may be illustrated by a statement of the penalty this imposes in a 10,000 ton armoured vessel. If at 10 knots this ship develops 1700 horse-power, there will be required at 15 knots, not one-third more, but 6200 horse-power—that is, over three times as much—and for 17 knots 12,000 horse-power, or an increase of 10,300 must be developed. This also demonstrates how much the ratio between speed and power falls; because if at 2000 horse-power 2.3 knots are gained for an increase of 1000 horse-power, at 12,000 for a similar increment of 1000 only one-quarter of a knot is obtained. In 1830 the steam pressure carried was from two to three pounds, and the coal expenditure each hour for every horse-power reached nine pounds; in 1886 the pressure had increased to 150 pounds, and the fuel consumption had fallen to 1.5 pounds, and today pressures of 200 pounds are to be utilized. As the swifter vessel with the higher economy is enabled to choose its range and position, and keep the sea for longer periods, it is easily seen that this question of speed is universally accepted as vital.

A parliamentary statement made in February shows that the following additions to the English fleet will be passed this year into the first-class reserve, and held ready for sea service at forty-eight hours' notice.

Thick armor battle-ship (*Hero*)	1
Partially armored ships of the *Admiral* class (*Rodney*, *Howe*, and *Benbow*)	3
Partially armored cruisers (*Warspite*, *Orlando*, *Narcissus*, *Australia*, *Galatea*, and *Undaunted*)	6
Partially protected cruisers (*Severn* and *Thames*)	2
Torpedo cruisers—six of the *Archer* class, one of the *Scout* class (*Fearless*)	7
Torpedo gun-boats of the *Rattlesnake* class	3
Composite gun-boats and sloops of the *Buzzard* and *Rattler* class	3
Total	25

At the end of 1887-88 one armoured ship, the *Camperdown*, and one protected cruiser, the *Forth*, will be nearly finished, the *Anson* will be approaching completion, and the new belted cruisers of the *Orlando* class will be far advanced. The armoured battleships *Victoria* and *Sanspareil*, of 10,470 tons displacement, are to be delivered according to contract in October, 1888, and the *Trafalgar* and *Nile*, the largest war-vessels yet laid down in England, are being pushed rapidly. Out of the thirty-seven ships building or incomplete at the commencement of 1887-88, twenty-six will be completed by the end of the year, thus leaving only nine of those specified, and two others not ordered, in the programme of 1885 to be finished subsequently. The ships projected for this year include:

20-knot steel-bottomed partially protected cruisers (*Medea, Medusa*)	2
19¾-knot copper-bottomed partially protected cruisers (*Melpomene, Marathon, Magicienne*)	3
Composite sloops of the *Buzzard* class (*Nymphe, Daphné*)	2
Composite gun-boats, improved *Rattlers*, (*Pigmy, Pheasant, Partridge, Plover, Pigeon, Peacock*)	6
Torpedo gun-boat of the *Grasshopper* class (*Sharpshooter*)	1
Total	14

Besides the ships that have been or will be finished in 1886-87, it is believed that thirty-five of the fifty-five first-class torpedo-boats (125 to 150 feet in length) will be added to the twenty which were completed in June.

In addition to the vessels mentioned above there are others not described, nor noticed in the text. The two battleships referred to earlier are the *Sanspareil* and the *Victoria*, the latter formerly known as the *Renown*, but named anew in April last. These ships are to carry 1180 tons of coal, and under forced draught are expected to develop 12,000 horse-power and a speed of 16.75 knots. The 9.2-inch 18-ton stern pivot gun originally intended for these vessels has been replaced by a 10-inch 26-ton rifle, and the secondary battery now includes twenty-one 6 and 3-pounder rapid fire guns. The other prominent departures from the central citadel type are the Nile and Trafalgar. These 11,940-ton ships are the largest war machines ever laid down for the British service. They are to carry revolving turrets on the fore and aft line amidships, and will have an intermediate broadside battery mounted in a superstructure which covers the full width of the ship between the turrets. A water-belt line 230 feet in length rises in the waist for a distance of 193 feet, and both belt and citadel are covered by a three-inch steel deck, which is curved forward and aft to strengthen the ram and protect the steering gear. The armour is compound—eighteen inches thick on the turret and twenty inches thick as a maximum on the water-line—and to support the backing there is an inner skin two inches thick. The armament consists of four 13½-inch 68-ton breech-loading rifles, two in each turret; of eight 5-inch guns in broad-side on a covered deck protected by three inches of vertical armour, and of eight 6-pounder rapid fire, ten 3-pounder Maxim, and four Gardner guns. The horse-power under forced draft is to be 12,000, and the estimated speed is 16½ knots. The main battery originally designed for these ships included only one 68-ton breech-loading rifle for each turret; subsequently this plan was rejected, and the armament stated above was adopted.

The economy of mounting the heavy guns in pairs arises not only from the increased power thus obtained from a given weight of guns, but from the fact that it requires but little more armour to protect two guns than to protect one. It also requires more machinery to work two guns separately than in pairs, and the magazine and ammunition supply arrangements of guns mounted separately are necessarily more complicated, and require more men to operate them than those mounted in pairs.

The French idea in mounting their heavy guns singly in three or four armoured barbettes is evidently so to distribute the gun-power as to leave a reserve of heavy guns in event of damage to one or more. But the demands for economy in weight are so great that two armoured structures widely separated would seem to furnish as satisfactory a scattering of the heavy gun-power as is justifiable. Guns mounted on the middle fine suffer less disturbance in rolling than those mounted either in the waist or *en echelon*, and their fire should be correspondingly more accurate. (*Recent Naval Progress*, June, 1887.)

The *Impérieuse* and *Warspite* have powerful ram bows, a steel protective deck, and a belt of compound armour which is 139 feet in length on the water-fine, 8 feet in width, and 10 inches thick. The engines were designed to develop 7500 horse-power and a speed of 16 knots, but on her trial the *Impérieuse* attained with forced draft a maximum speed of 18.2 knots and 10,344 horse-power, and a mean speed, after four runs on the measured mile, of 17.21 knots. In September, 1886, with all guns and stores in place, and with 900 tons of coal in the bunkers, the *Impérieuse* developed a mean speed of 16 knots. The armament is composed of four 9.2-inch guns, mounted in four 8-inch plated circular barbettes, and situated one forward, one aft, and two in the waist; on the gun-deck there are six 6-inch guns, and the secondary battery is made up of twelve 6-pounder rapid fire, ten 1-inch Nordenfelt, and four Gardner guns, and of four above-water and two submerged torpedo tubes. Owing to the increased weights of the armament, stores, machinery, and equipments put in these vessels since they were first designed, the draught of water is now found to be nearly three feet greater than was intended. It is only fair to state that they were originally expected to carry but 400 tons of coal, though curiously enough, when this fuel capacity was subsequently increased to 1200 tons, no allowance was made for the additional armoured surface required.

The armoured free-board was to have been 3 feet 3 inches at a draught of 25 feet, but the supplementary weights increased the draught Ill-inches and reduced this free-board to 2 feet 3½ inches; and later, when the full bunker capacity of 900 tons was utilized, the draught was again increased 14 inches, and the free-board lowered to 1 foot 1½ inches. Finally it was for a time determined to carry 1200 tons of coal, though this would result, when the ship was fully equipped for sea, in bringing the top of the armoured belt nearly flush with the water.

> As four of the torpedo tubes are above water, and have ports cut through the armour belt, this decrease of free-board rendered them useless, it having been shown during an experimental cruise on the *Impérieuse* in December, 1886, with but 800 tons of coal on board and in a calm sea, that in attempting to discharge the broadside torpedoes they jammed in the tubes, and altered shape to a dangerous degree. In order to make them of any use they will have to be restored to their intended height above the water-line. It is believed this can be accomplished by removing part of the superstructure, by dispensing with all top-hamper and its attendant supply of stores, equipments, etc., and by limiting the maximum coal supply to the bunker capacity of 900 tons. The masts are accordingly being removed from both vessels, leaving them but one signal mast stepped between the funnels, and fitted with a military top. (Lieutenant Colwell, U.S.N., in *Recent Naval Progress*, 1887.)

In May, 1886, the *Warspite*, when very light, developed with natural draft 7451 horse-power and a speed of 15½-knots on a consumption of 2.69 pounds of coal each hour per horse-power; and with forced draft 10,242 horse-power and a speed of 17¼ knots were obtained on a similar consumption of 2.9 pounds of coal.

In the minority report of the 1871 Committee on Designs Admiral Elliot and Rear Admiral Ryder...

> ... strongly advocated the use of a protective deck in conjunction with other features, instead of side-armour, for protection to stability. The idea as regards cruisers was first carried out in

the full-rigged ships of the English *Comus* class of 2380 tons displacement and 13 knots speed, launched in 1878, in which the engines, boilers, and magazines were covered by a horizontal 1½-inch steel deck placed below the water-line, the space immediately above containing cellular subdivisions.

Then followed, in 1882, the *Leander* and her three sister bark-rigged vessels, which are a compromise between the speed of the *Iris* and the protection of the *Comus*. They are of 3750 tons displacement and 17 knots maximum speed; they carry ten 6-inch 4-ton B. L. R., and 725 tons of coal, and have a 'partial protective deck,' covering engines, boilers, and magazines, which is 1½ inches thick, and which bends down below the load water-line at the sides. Our new cruisers, the *Chicago, Boston*, and *Atlanta*, bear a closer resemblance to this type than they do to any other in respect of their protection. About this time the Chilean cruiser *Esmeralda*, of 3000 tons, appeared, having a protective deck complete from stem to stern-post, carrying an exceptionally heavy battery and coal supply, and withal attaining the unprecedented speed of 18.28 knots. Italy was not slow to perceive the advantages of this type, and accordingly bought an improved *Esmeralda*, the *Giovanni Bausan*, and at once commenced to build four others, the *Vesuvio, Stromboli, Etna*, and *Fieramosca*, each of 3530 tons. Japan ordered two improved *Esmeraldas*, the sister ships *Naniwa-Kan* and *Tacachi-ho-Kan*, from Armstrong, in England, and a similar vessel, the *Unebi*, in France, while England laid down a similar class, the *Mersey* and three others, and France a similar cruiser, the *Sfax*, of 4400 tons. (Lieutenant Chambers, U.S.N.)

The *Unebi* was a bark-rigged, twin-screw, protected steel cruiser of 3651 tons. Her armament consisted of four 9.45-inch breech-loaders on sponsons, six 5.9-inch breech-loaders in broadside, one 5.9-inch bow pivot, twelve rapid fire and two Nordenfeldt machine guns, and a supply of Whitehead torpedoes. In September, 1886, she developed with forced draft 7000 horse-power and an average speed of 18.5 knots during four runs over the measured mile. She sailed for Japan in November, 1886, with a French crew numbering seventy-eight men, left Singapore for Yokohama on December 3, 1886, and has never been seen nor heard of since. She is said to have been top-heavy, and to have rolled dangerously in a sea way.

The *Naniwa-Kan,* a steel cruiser, 300 feet in length and 46 feet in beam, has on an extreme draught of 19 feet 6 inches a displacement of 3730 tons; the steel hull is fitted with a double bottom under the engines and boilers, and has a strong protective deck, two to three inches thick, which extends from the ram to the stern-post, and carries its edges four feet below, and its crown one inch above, the load water-line. There are ten complete transverse and several partial water-tight bulk-heads; the space between the protective and the main deck is minutely subdivided into compartments, which are utilized as coal-bunkers, store-rooms, chain-lockers, and torpedo-rooms; the conning tower is protected by two inches of steel armour; and two ammunition hoists, three inches thick, lead from the shell-rooms to the loading towers at the breech of the two heavy guns. The armament consists of two 10-inch 28-ton breech-loading rifles on central pivots, with 2-inch steel screens, and of six 6-inch guns, with a secondary battery of two 6-pounder rapid fire, eight 1-inch Nordenfeldt, four Gardner guns, and four above-water torpedo tubes. The engines are of the horizontal compound type, situated in two compartments, one abaft the other, and there are six single-ended locomotive three-furnace boilers in two separate compartments, with athwartships fire-rooms; the indicated horse-power under a forced draft was 7650, and the maximum speed 18.9 knots. This has since been exceeded. The *Mersey* and her class—the *Severn*, *Thames*, and *Forth*—like the *Naniwa* are unarmoured steel cruisers, with a complete protective deck, the horizontal portion of which is one foot above, and the inclined three inches below, the water-line. The main battery of these ships consists of two 8-inch guns, mounted on central pivots forward and abaft a covered deck which carries ten 6-inch guns; the secondary battery has ten 1-inch Nordenfeldt and two Gardner machine guns, and there are six above-water torpedo tubes in broadside.

> The development of the *Mersey* design has resulted in the new English 'belted cruisers,' in which, to satisfy the demand for a water-fine belt of armour, the displacement has been increased to 5000 tons.

> The five originally projected—the *Orlando*, *Narcissus*, *Australia*, *Galatea*, and *Undaunted*—together with the *Immortalité*, subsequently laid down, have already been launched, and an additional cruiser of the same type, the *Aurora*, is well advanced. The general construction is similar to the Naniwa and Mersey, the larger tonnage being given in order to carry a water-line belt, which is ten inches thick, stretches for 190 feet amidships, and was intended to extend from 1½ feet above to

four feet below the load water-line. The armoured deck is from two to three inches thick, and the conning tower is thirteen inches. The triple expansion engines are planned to develop 8500 horse-power and a speed of 18 knots. Like the *Impérieuse* and *Warspite*, these vessels are found to draw much more water than was originally proposed. When designed in 1884 they were expected to have, with all weights on board, a mean draught of twenty-one feet, and to carry above water eighteen inches of the five feet six inch armour-belt. But a fever for improvement set in so valorously that the changes made in armament and machinery added one hundred and eighty-six tons to the displacement and increased the draught seven inches—that is, an amount which left the top of the protective belt only eleven inches above the smooth water-line. This submersion did not, however, cool the ardour of the Admiralty officials, for it has been decided that the nine hundred tons of coal originally fixed as the fuel supply must be carried; the immediate result of this is said to be an increase in the draught of eighteen inches, and a disappearance of the armour-belt to a point nearly six inches *below* the water-line. Subsequent improvements will be awaited with great interest, especially by those American journalists of inquiring tendencies who envyingly detect between the promise and performance of these ships opportunities which, had they occurred at home, would have enabled them to swamp our naval service and its administration in billows of pitiless ink.

The most popular naval event of the year was the review in July of the British fleet assembled at Spithead. The one hundred and twenty-eight war-vessels participating included three squadrons of armoured vessels and cruisers, aggregating thirty-four ships, seventy-five torpedo-boats and gun-boats, divided into five flotillas, six training brigs, and thirteen troop-ships. Besides these there were the troop-ships appointed to carry the distinguished visitors, and the small vessels and dockyard craft allotted to the corporation of Portsmouth.

The warships were drawn up in four lines, facing up channel, the starboard column lying opposite the Isle of Wight, and the port column off Portsmouth. The ships were two cables and the columns three cables apart. The flotillas were ranged in double columns between the port line of the armoured vessels and the main-land, and the troop-ships were placed in single column between the starboard line and the Isle of Wight. This made four lines of vessels on one side of the channel and three on the other, extending from South Sea Castle to the Rye Middle Shoals, or a distance of two miles. No such fleet was ever seen before in time of peace, for every class of the British navy was so well represented that the review of the Crimean fleet by the queen and the prince consort, thirty-one years ago, suffered by comparison. Some of the wooden ships which figured at that time were present, and the wide differences in everything bore strong testimony to the developments which have been made within a generation. Nelson's old ship, the *Victory*, was a conspicuous object, and her timbers echoed again and again with cheers as boat after boat passed her. More than that, the old ship mounted a gun or two and joined in the universal salute to the queen. Shortly after two o'clock the *Euphrates*, *Crocodile*, and *Malabar* hove to off Osborne as an escort to the royal yachts when the queen embarked.

The queen left Osborne House a few minutes before three o'clock, went aboard the royal yacht *Victoria and Albert*, and left the buoy in the bay promptly at the hour fixed. She was preceded by the *Trinity* yacht and followed by the royal yachts *Osborne* and *Alberta*, and by the war-vessels *Enchantress, Helicon, Euphrates, Crocodile,* and *Malabar*. The royal procession proceeded straight to its destination and passed between the lines, leaving the coast-defence ships, gun-boats, and torpedo-boats on the port hand. After steaming as far as the Horse Elbow buoy the *Victoria and Albert* turned to starboard, passed between the two columns of large ships, and then between the lines of the foreign war-vessels. As the yacht steamed slowly by the warships the crews cheered loudly, but it was not until the queen had gone through the double line that the royal salute was fired. On board such vessels as had no masts the turrets, breastworks, and decks were lined with the crews, and the spectacle was as splendid as it was potent with an earnest evidence of mighty power. Altogether the fleet extended over four miles, and even this length was added to by the great troop-ships which steamed into line and saluted the queen as she made her progress.

The jubilee week was not without its accidents, for the *Ajax* and *Devastation* collided at the rendezvous, and subsequently the *Agincourt* and *Black Prince* had a similar experience. These mishaps evoked much hostile criticism, and among other things gave currency to an ex-tract from a speech made by Lord Randolph Churchill several weeks before. Speaking of the navy, he had declared that:

> In the last twelve or thirteen years eighteen ships have been either completed or designed by the Admiralty to fulfil certain purposes, and on the strength of the Admiralty statements parliament has faithfully voted the money. The total amount which either has been or will be voted for these ships is about ten mil-

lions, and it is now discovered and officially acknowledged that in respect of the purposes for which these ships were designed, and for the purposes for which these ten millions either have been or will be spent, the whole of the money has been absolutely misapplied, utterly wasted and thrown away.

Sir Charles Dilke does not agree with this pessimism of his political opponent, though he, too, has something to say of the British fleet, in relation to its influence upon the present position of European politics, which is well worth quoting.

> There is less to be said in a hostile sense with regard to the present position of the navy, than may be said, or must be said, about the army. Clever German officers may write their 'Great Naval War of 1888,' and describe the destruction of the British fleet by the French torpedo-boats, but on the whole we are not ill-satisfied with the naval progress that has been made in the last three years. There is plenty of room for doubt as to whether we get full value for our money; but at all events our navy is undoubtedly and by universal admission the first navy in the world, and relatively to the French we appear to show of ships built and building a number proportionate to our expenditure. The discovery of the comparative uselessness of automatic torpedoes is an advantage to this country, and no great change in the opposite direction has recently occurred. M. Gabriel Charmes has pointed out to France the manner to destroy our sea-borne trade, but excellent steps have been taken since his book appeared to meet the danger which he obligingly made clear to us. It remains a puzzle to my civilian mind how Italy can manage to do all that in a naval sense she does for her comparatively small expenditure, and how, spending only from a fourth to a sixth what we spend upon our navy, she can nevertheless produce so noble a muster of great ships. But our naval dangers are, no doubt, dangers chiefly caused rather by military than by naval defects. Our navy is greatly weakened for the discharge of its proper duties by the fact that duties are thrown upon it which no navy can efficiently discharge. As Admiral Hoskins has said, it is the duty of the commander of the British Fleet to drive the hostile squadrons from the seas, and to shut up the enemy's ships in his different ports; but, on the other hand, he has a right to expect that our own ports and coaling stations shall be protected by batteries and by land

forces. This is exactly what has not yet been done, although the defence of our coaling stations by fortresses and by adequate garrisons is essential to the sustaining of our maritime supremacy in time of war.

It is only, however, by comparison with our army that I think our navy in a sound position. In other words, our military situation is so alarming that it is for a time desirable to concentrate our attention upon that, rather than upon the less pressing question of the condition of the navy. I must not be thought, however, to admit, for one single instant, that our navy should give us no anxiety. As long as France remains at peace, and spends upon her navy such enormous sums as she has been spending during the last few years, she will be sufficiently near to us in naval power to make our position somewhat doubtful; make it depend, that is, upon how the different new inventions may turn out in time of war. Our navy is certainly none too large (even when the coaling stations and commercial ports have been fortified, and made for the first time a source of strength rather than of weakness to the navy) for the duties which it has to perform. It would be as idle for us, with our present naval force, to hope to thoroughly command the Mediterranean and the Red Sea against the French without an Italian alliance, as to try to hold our own in Turkey or in Belgium with our present army. Just as the country seems now to have made up its mind to abandon not only the defence of Turkey against Russia, but also the defence of the neutrality of Belgium, so it will have to make up its mind, unless it is prepared to increase the navy, to resort only to the Cape route in time of war. Italy being neutral, and we at war with France, we could not at present hope to defend the whole of our colonies and trade against attack, and London against invasion, and yet to so guard the Mediterranean and the Red Sea as to make passage past Toulon and Algiers, Corsica and Biserta, safe. Our force is probably so superior to the French as to enable us to shut up their ironclads; but it would probably be easier to shut in their Mediterranean ironclads by holding the Straits of Gibraltar than to attempt to blockade them in Toulon. I confess that I cannot understand those Jingoes who think that it is enough to shriek for Egypt, without seeing that Egypt cannot be held in time of war, or the Suez route made use of with the military and naval forces that we possess at present.

As against a French and Russian combination of course we are weaker still. Englishmen are hardly aware of the strength of Russia in the Pacific, where, if we are to attack at all, we must inevitably fight her, and where, if we are to adopt the hopeless policy of remaining only on the defensive, we shall still have to meet her for the protection of our own possessions. Just as the reduction of the horse artillery, comparatively unimportant in itself, has shown that the idea of the protection of Belgian neutrality has been completely given up, so the abandonment of Port Hamilton, instead of its fortification as a protection for our navy, seems to show that we have lost all hope of being able to hold our own against Russia in the North Pacific. On the 1st of August Russia will have upon her North Pacific station—cruising, that is, between Vladivostock and Yokohama—three new second-class protected ships—the *Vladimir, Monomakh*, and the *Dmitri Donsköi*, of nearly six thousand tons apiece, and the *Duke of Edinburgh*, of four thousand six hundred tons; one older protected ship, the *Vitiaz*, of three thousand tons; four fast-sailing cruisers—the *Naïezdnik,* the *Razboïnik*, the *Opritchnik*, and the *Djighite*; and four gun-boats, of which two are brand-new this year, (as at time of first publication). While talking about their European fleets, the Russians are paying no real attention to them, and are more and more concentrating their strength in the North Pacific.

Cusack Smith, *Our War Ships*, says:

The British navy is not in danger, and the British navy, whatever its shortcomings, is relatively far stronger than its thoughtless detractors would have us believe. Our ships do steer and our ships do steam—at least as well as those of other powers; and, what is more, our ships will 'fight' and our ships will 'win,' in spite of the dismal forebodings of interested panic-mongers.

With the resources at our command, our armaments afloat admit of a rapid development, in which no other country can compete with us. A French writer has truly said, '*La puissance d'une marine est moins dans son matériel à flot, que dans l'outillage de ses arsenaux, et dans la puissance productive de ses chantiers.*'

As a maritime power we are unequalled, and if we be true to ourselves we shall remain so.

In 1886 the fighting ships of the British navy were summarized as follows:

Ships.	No.	Guns.	Displacement.	Horse power.
ARMORED.			Tons.	
In commission and in reserve	50	508	339,750	241,390
Deduct ships of doubtful value	7	137	50,780	30,970
Total reliable armored ships	43	371	288,970	210,420
UNARMORED.				
In commission and in reserve	197	1121	221,957	245,692
Deduct ships of doubtful value	15	76	27,760	27,470
Total reliable unarmored ships	182	1045	194,197	218,222
Armored ships building	12	148	89,660	114,000
Unarmored ships building	21	112	24,650	53,250
Total	33	260	114,310	167,250
Armored ships being completed	10	93	84,880	84,750
Unarmored ships being completed	10	90	26,790	41,800
Total	20	183	111,670	126,550
Total armored ships	72	749	514,290	440,140
Total unarmored ships	228	1323	273,397	340,742
Grand total of ships	300	2072	787,687	780,882

During the last year thirty-seven vessels of the following classes were stricken from the list, *viz.*, five armoured ships, seven cruisers of the third class, sixteen gun-vessels, one despatch-boat, and eighteen special service gun-boats.

The following table shows the armoured and partially protected ships now under construction or lately finished:

Name of Ship.	Tons.	Horse-power.	Speed.	Total Cost.	Armament.
TURRETS.					
Trafalgar	11,940	12,000	16.5	£844,318	4 67-ton, 8 5-in.
Nile	11,940	12,000	15.5	889,421	4 67-ton, 8 5-in.
Victoria	11,470	12,000	16.75	829,979	2 110-ton.
Sanspareil	11,470	12,000	16.75	825,468	1 10-in., 12 6-in.
Edinburgh	9,150	7,500	15.4	683,609	4 45-ton, 5 6-in.
Hero	6,200	6,000	15.5	421,500	2 45-ton, 4 6-in.
BARBETTES.					
Anson	10,000	12,500	17.5	752,288	4 67-ton, 6 6-in.
Camperdown	10,000	11,700	17.5	743,074	4 67-ton, 6 6-in.
Benbow	10,000	10,850	17.5	810,633	2 110-ton, 10 6-in.
Howe	9,700	11,500	17.0	720,771	4 67-ton, 6 6-in.
Rodney	9,700	11,500	17.0	726,482	4 69-ton, 6 6-in.
Collingwood	9,150	9,570	16.5	670,752	4 44-ton, 6 6-in.
Impérieuse	8,500	10,344	17.21	559,901	4 9.2-in., 6 6-in.
Warspite	8,500	10,243	17.25	558,449	4 22-ton, 6 6-in.
BELTED CRUISERS.					
Immortalité	5,000	8,500	18.0	302,920	2 22-ton, 10 6-in.
Aurora	5,000	8,500	18.0	308,585	2 22-ton, 10 6-in.
Australia	5,000	8,500	18.0	290,613	2 9.2-in., 10 6-in.
Galatea	5,000	8,500	18.0	290,300	2 9.2-in., 10 6-in.
Narcissus	5,000	8,500	18.0	290,751	2 9.2-in., 10 6-in.
Orlando	5,000	8,500	18.0	299,905	2 9.2-in., 10 6-in.
Undaunted	5,000	8,500	18.0	299,525	2 9.2-in., 10 6-in.
PARTIALLY PROTECTED CRUISERS.					
Mersey	3,500	6,000	18.0	236,435	2 8-in., 10 6-in.
Severn	3,500	6,000	18.0	234,282	2 8-in., 10 6-in.
Thames	3,500	5,700	18.0	227,980	2 8-in., 10 6-in.
Forth	3,500	5,700	18.0	221,918	2 8-in., 10 6-in.

The French Navy
by Sir Edward J. Reed

We have now to pass under review that vast array of naval constructions which the Continental navies of Europe offer to our observation. It is not at all surprising that the introduction of steam-engines, of iron and steel hulls, and of armour-plating has been attended throughout Europe by even greater diversity of thought and practice than has characterized our naval progress—"our progress" here signifying that of both the United States and Great Britain. And this may, I think, truthfully be said without in any degree neglecting the striking originality of the American Monitors, to which I endeavoured to do justice.

As regards two of the three great changes just adverted to, the only differences of opinion that have arisen have been in the nature of competitions rather than of conflicts. No one, so far as I am aware, has ever proposed to revert to sail-power or to wooden hulls in important ships-of-war. On the contrary, the powers have been in continual competition in the effort to reduce the weights of the hulls of warships (apart from armour) by the extended use, first of iron, and afterwards of steel, and to apply the savings of weight thus effected to the development of engine-power, speed, and steaming endurance. On the other hand, it must be acknowledged that the development of armour has been pursued with less constancy and less earnestness, the result being that marked contrasts are exhibited by European navies.

It may be said, with little or no qualification, that all other European naval powers followed, in the first place, the example set by the late Emperor Napoleon III., in *La Gloire*, by covering the whole of the exposed part of the warship's hull with armour-plating. All the early ironclads of Russia, Italy, Austria, and Germany were protected from stem to stern, and from a few feet below water to the upper deck. England did the same in the cases of a few ships, although she began, as we saw before, with the *Warrior* type, in which the armour was limited to the central part of the ship. But the system of completely covering the exposed ship with armour has now entirely and properly, passed away from European practice, and has been succeeded by varied arrangements of armour.

The importance of giving effectual protection to the hull "between wind and water," as it is called (signifying from a few feet below the water-line to a few feet above that hue), has been steadily recognized by Continental governments, with but the rarest exceptions. Nothing corresponding to that wholesale abandonment of armour for about a hundred feet at each end of the ship which has been practised in the British ships of the *Inflexible* and *Admiral* types is displayed in the line-of-battle ships of the Continent. In France, indeed, two such ships were laid down under some temporary influence, *viz.*, the *Brennus* and the *Charles Martel*, but they appear to have soon fallen under suspicion, and there has not been, to my knowledge, any great disposition to complete them for service. A return made by the Admiralty to the order of the House of Commons has been printed, and says of the *Brennus* and *Charles Martel*:

> Though these vessels still appear in the list of the French Navy, but little money has been voted for their construction in 1886, and all work on them is now reported to have been stopped.

I know not what significance is to be attached to the fact, but I observe that these two ships were omitted altogether from the iron-clad ships of France published so recently as May, 1886, in the *Universal Register* of shipping, which Lloyd's Register committee "believe will be found the most complete list that has yet been published."

It seems not improbable, therefore, that the dangerous system of exposing two-thirds of the ship's length to destruction from all kinds and every system of naval guns, even the smallest, which prevailed in the British navy for more than twelve years, and which has now happily been superseded in the powerful new ships *Nile* and *Trafalgar*, obtained but little more than momentary approval in France, and is likely to have led to the condemnation of the only two ships in which it was attempted—a result which is creditable alike to French science and to French sagacity.

In Italy the *Inflexible* system (which has met in France with the fate we have just seen) obtained temporary favour, and was adopted in the

Duilio and the *Dandolo*, two very large ships, of 11,000 tons each, of a speed exceeding fifteen knots, and each carrying four 100-ton guns in turrets. Although these ships are 340 feet in length, even the armoured belt amidships (if "belt" in any sense so short a strip of armour may be called) is but 107 feet long, leaving therefore 233 feet of the ship at the ends wholly devoid of water-line protection. As the author of the "citadel system," I cannot regard such an arrangement as this as a fair and reasonable embodiment of it, the discrepancy between the armoured and unarmoured portions being greater in these two ships than even in the *Ajax* and *Agamemnon*, which are perhaps the worst examples of the abuse of the citadel system in the British navy. It is to the credit of the Italian government that ships of this type were not repeated in their navy; and it is but right to point out that there were excuses (which probably ranked in the minds of the designers as *reasons*) for a more extreme proportionate limitation of the citadels being adopted in the *Duilio* and *Dandolo* than in the *Ajax* and *Agamemnon*. Among these were the possession by the Italian ships of heavier armaments, and of far greater steam-power and speed than the British ships possessed—a matter to which further reference will be made hereafter—and probably, also, the adoption of somewhat finer water-lines as a means of attaining the superior speed.

In this connection it may be well to observe that the question of leaving so-called armoured line-of-battle ships without armour at the extremities is first one of principle, and afterwards one of degree. The principle (which should be observed in the design of every armoured vessel which is intended for the line of battle, or for those close and severe contests of ship with ship which will probably supersede in a great degree the system of fighting in fines of battle) is this: the proportion which the armoured citadel bears to the unarmoured ends must always be such as to enable the ship to keep afloat all the time the armour itself holds out against the attack of the enemy; so that injuries to the unarmoured ends, however great or multiplied, shall not alone suffice to destroy the ship. Whatever may occur in the future to interfere with the application of this principle—and I do not deny that such interferences may arise under certain perfectly conceivable circumstances—nothing has yet happened to justify its abandonment, or to even justify the remotest chance of its being violated.

If a ship is not intended to close with an enemy, or to fight her anyhow and anywhere on the open sea—which certainly has been the dominant idea of the British navy, in so far as its great line-of-battle ships are concerned—if, for example, a combination of immense speed with one or two extremely powerful and well-protected guns should serve a particular object better than a slower and more fully protected ship would serve it—then even great destructibility in the ship itself may justifiably be incurred. But for general naval service, and in every case in which a ship is intended to accept battle with a powerful antagonist and fight it out, or to force an action when she encounters such an enemy, it cannot be wise to leave her so exposed that that enemy may almost certainly sink her or cause her to capsize by merely pouring any kind of shot or shell into her unarmoured parts. But even the observance of the above general principle is not alone all that is desirable in armoured line-of-battle ships. It is not well to leave even so much of the ends of such ships wholly exposed as may lead to the speedy loss in action of her steaming or steering powers. The armour-belt should be of sufficient length to fairly guarantee the ship against prompt disablement in action, and to do this it must be carried very much nearer to the bow and stern than it has been in the cases of the Italian ships (*Duilio* and *Dandolo*) now under notice.

On the other hand, where ships are formed with fine water-lines, and the two opposite sides are consequently very near to each other for many feet, it is quite unnecessary to cover them with armour. The buoyancy comprised between the two sides at such parts is very small, and consequently penetration can let but little water into the ship, and do but little harm. It is a matter for the exercise of professional judgement where to draw the line between the armoured and the unarmoured parts. In the new British ships *Nile* and *Trafalgar*, which have excited great admiration in England, there are about sixty feet of length at each end left without armour, and as the ships have tine lines, but are nevertheless of considerable breadth at sixty feet from the ends, it seems probable that good judgement has been shown by their designers in this matter.

I have discussed this question at some length because it is one of primary consideration in the design of important armoured ships, and because the abandonment of a long belt of armour is also one of the few features of construction respecting which the designers of the Continent have steadfastly refrained from following the example set by the Admiralty Office at Whitehall from the years 1870 to 1885. It will complete the consideration of this branch of the subject to say that there are numerous ships of the ironclad type in foreign navies in which the armour (justifiably, as has just been shown) stops somewhat short of the ends, but very few indeed in which the length of the unarmoured parts exceeds that of the armoured. Among the last named may be mentioned a very questionable class of vessels (*Sachsen* type) in the German navy, and a much smaller sea-going vessel belonging to the Argentine Republic, named the *Almirante Brown*, which is a well-designed vessel in other respects, but which, on account of her long defenceless bow and stern, would do better to avoid than to fight an enemy.

It will be instructive to repeat here, before leaving this question of partially armoured ships, a comparison resembling that which I employed in a paper read at the Royal United Service Institution, in which are set down in one column the displacements of certain British and French ships, eleven of each, built and building, possessing maximum armour on the water-line of at least fifteen inches. As all the French ships given have complete or all but complete armour-belts, it is proper to reckon their whole displacement tonnages as armoured tonnage. But in the case of all the British ships which carry such thick armour they are deprived of armour altogether except amidships, and it is therefore misleading, and even absurd, to reckon their whole displacement tonnages as armoured tonnage. For this reason I am obliged to give two tonnages for them, *viz.*, the armoured and the unarmoured, as I do below:

French Ships.	Armored.	British Ships.	Unarmored.	Armored.	Total.
	Tons.		Tons.	Tons.	Tons.
Amiral Baudin	11,141	Inflexible	5,210	6,670	11,880
Amiral Duperré	10,486	Ajax	4,160	4,350	8,510
Dévastation	9,639	Agamemnon	4,160	4,350	8,510
Formidable	11,441	Colossus	4,580	4,570	9,150
Courbet	9,639	Edinburgh	4,580	4,570	9,150
Hoche	9,864	Collingwood	4,580	4,570	9,150
Magenta	9,864	Rodney	4,800	4,900	9,700
Marceau	9,864	Howe	4,800	4,900	9,700
Neptune	9,864	Camperdown	4,900	5,100	10,000
Caïman	7,239	Benbow	4,900	5,100	10,000
Indomptable	7,184	Anson	4,900	5,100	10,000
Total	106,225	Total	51,570	54,180	105,750

I have not thought it necessary to alter these figures in repealing this comparison, as they are sufficiently near the truth for the only purpose for which I employ them, which is that of exhibiting the fact that whereas the above eleven British ironclads (so called) figure in the official tables of the British government as constituting an armoured tonnage of 105,750 tons, nearly equal to that of the eleven French ships, they really represent but little more than half that amount of armoured tonnage.

Having now dealt with the primary question of the defence of ships by means of armour-belts, we come to the greater or less de-

THE DÉVASTATION: FRENCH ARMOURED SHIP OF THE FIRST CLASS

fence bestowed upon them above water. The course taken by the French designers, when the increased thickness of armour made it impossible to repeat the complete protection adopted in *La Gloire* and her compeers, was in some few cases that of belting the ship with armour, and giving great "tumble home" to the sides above water, excepting at the central armoured battery, thus allowing that battery to project, and its guns to fire directly ahead and astern, past the inwardly inclined sides. This system has been strikingly carried out in the two sister ships *Courbet* and *Dévastation*, the former of which is shown, stem on, in the cut on a following page, which is engraved from a photograph taken after her launch, and before she began to receive her armour-plating. A representation of the sister vessel, *Dévastation* (forming one of the series of engravings given in this chapter from drawings specially executed for the purpose by Chevalier De Martino), forms our following illustration.

But generally in the French navy, and in nearly all but its earliest ships, direct head and stern fire has been obtained by means of elevated and projecting towers, armour-plated to a sufficient height to protect the gun machinery, but with the guns themselves unprotected, and firing *en barbette*. In the case of the two ships *Dévastation* and *Courbet* the main-deck projecting battery carries four guns, each commanding a full quadrant of a circle. The barbette batteries, standing up above the upper deck, carry a powerful gun on each side of the ship, with great range of fire.

Having given these general indications of the system of attack and defence adopted in the French navy—by far the most important of all the Continental navies—it now becomes desirable to go more into particulars. It is not necessary to dwell upon the early ironclads of France. The *Gloire* and a dozen others of like character were all built of wood, without water-tight bulkheads, without rams or spurs, with armour-plates from four to six inches thick only, and with guns of small calibre and power. They may be left out of consideration in dealing with the present French navy. They were followed by six other vessels, also built of wood, but with upper works of iron, *viz.*, the *Océan, Marengo, Suffren, Richelieu, Colbert, Trident*. They were armoured with plates of a maximum thickness of 8½ inches, and carried four guns of 10¾ inches calibre, weighing 23 tons each, with four 16-ton guns, and half a dozen light ones. They varied in some particulars, ranging in tonnage from 7000 to 8000 tons, in horse-power from 3600 to 4600, and in speed from 13 to 14½ knots. The *Friedland* is another vessel which is frequently classed with the previ-

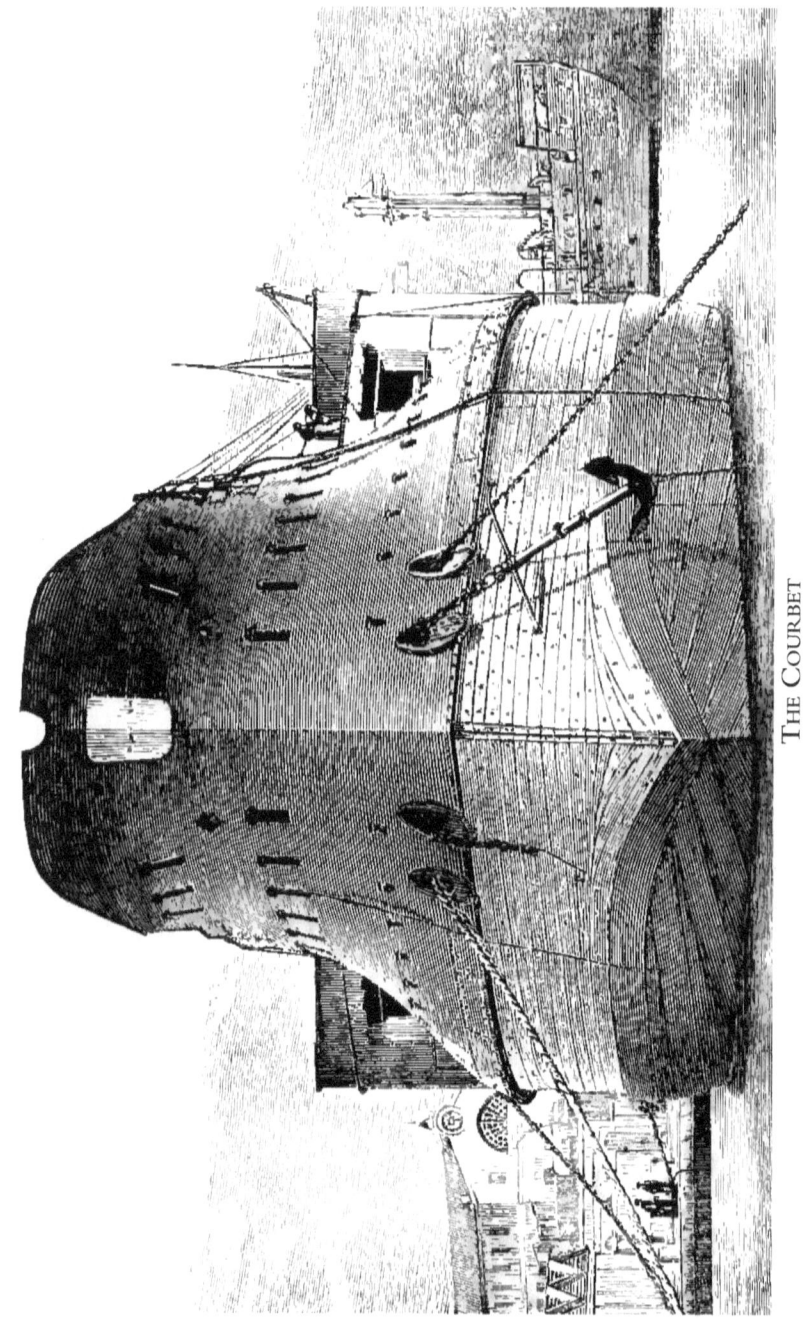

THE COURBET

ous six ships, the largest of which she generally resembles, but she is built of iron, and carries eight 23-ton guns, and none of the 16-ton. A committee which sat in 1879, and which had for its president and vice-presidents men no less eminent than the late M. Gambetta and MM. Albert Grévy and Jules Ferry, pronounced these seven ships to be the strongest armoured ships of the French navy then in service. Such great advances have since been made, however, that it is only necessary to add respecting these vessels that they were nearly all single-screw ships, and that they carried their principal armament at broadside ports on the main-deck, and in raised barbette towers placed at the four corners of the central battery. The *Richelieu* was the largest of these vessels.

Not one of the foregoing French ships of the early period conformed to conditions which were laid down officially in 1872 as those requisite for first-class French ironclads, *viz.*, that they should be constructed of iron (or steel), with water-tight compartments, be armoured with plates 12 inches thick, with decks from 2 to 2½ inches thick, armed with guns of 24 centimetres calibre, commanding certain prescribed ranges of fire, and furnished with spurs or ram stems. There were, however, four ships then under construction or trial which did conform to the prescribed conditions, *viz.*, the two already spoken of—the *Courbet* and *Dévastation*, and two others named the *Redoutable* and the *Amiral Duperré*. With these powerful ships may be said to have commenced the era of iron and steel line-of-battle ships in France. We will now bring them, together with still more recent French ships of the first class, into a table in which their particulars may be conveniently grouped.

TABLE A.—MODERN FRENCH ARMORED SHIPS OF THE FIRST CLASS.

Name of Ship.	Displacement in tons.	Indicated Horsepower.†	Speed in Knots.†	Length.	Breadth.	Draught of Water.	Maximum Thickness of Armor.	Heaviest Guns carried.
				Feet.	Feet.	Feet.	Inches.	
Amiral Baudin..	11,200	8,320	15	319	70	25.8	22	3 of 75 tons.
Amiral Duperré.	10,800	8,120	14.2	319	70	25.8	22	4 " 48 "
Dévastation	9,900	8,320	14.5	312	69.8	25.5	15	{4 " 48 " / 4 " 28 "}
Formidable	11,260	8,320	15	319	70	25.8	22	3 " 75 "
Foudroyant (now Courbet)	9,500	8,200	15	311	69.8	25.5	15	{4 " 48 " / 4 " 28 "}
Hoche..........	10,480	5,500	14	329	66	26.5	17.7	4 " 52 "
Magenta........	10,480	5,500	14	329	66	26.5	17.7	4 " 52 "
Marceau........	10,480	5,500	14	329	66	26.5	17.7	4 " 52 "
Neptune........	10,480	5,500	14	329	66	26.5	17.7	4 " 52 "
Redoutable.....	9,080	6,000	14.2	312	64.6	24.4	14	{4 " 28 " / 4 " 24 "}
Caïman.........	7,200	4,800	14	271	59	23	17.5	2 " 48 "
Furieux.........	5,700	3,400	12	248	59	21.4	17.5	2 " 48 "
Indomptable....	7,200	4,800	14	271	59	22.8	19.5	2 " 75 "
Requin.........	7,200	6,000	14.5	271	59	22.8	19.5	2 " 75 "
Terrible........	7,200	4,800	14	271	59	22.8	19.5	2 " 75 "
Tonnant........	4,707	1,750	10	248	58.4	17.8	17.5	2 " 48 "

For the reason before stated, the *Brennus* and *Charles Martel* are omitted from this table. Indicated horse-power and speed in knots are taken from *Lloyd's Universal Register*.

The ship which alphabetically falls last in this table among the ships of 9000 tons and upwards, the *Redoutable*, came first in point of time, *viz.*, in 1872, and her design marked the commencement of the new era in French ironclad construction. One of the features of the change was, as already intimated, the abandonment of wooden hulls, which we had succeeded in accomplishing in England eight years before. The first design proposed by myself to the British Admiralty provided for an iron hull, and although the force of circumstances compelled us to construct my earliest war-vessels in timber, yet so strongly averse were we to the employment of so perishable a material as wood within an iron casing that Admiral Sir R. Spencer Robinson succeeded in preventing the construction of three out of five wooden line-of-battle armoured ships that had previously been proposed by the government of the day, and sanctioned by parliament. This was in 1863 or 1864, the *Lord Clyde* and *Lord Warden* being the last large armoured wooden ships laid down in her Majesty's dockyards.

The French delayed the change for some years, as we see. M. De Bussy, the designer of the *Redoutable*, and a most accomplished naval constructor, built a very large part of the ship of steel, and by so doing brought the French dockyards into early acquaintance with the superiority of that material to iron for constructive purposes. The *Redoutable* has armour of more than 14 inches in thickness upon her belt, and of 9½ inches upon her central battery. She carries eight 25-ton guns—some returns say four of 28 tons, and four of 24 tons, all being of 27 centimetres calibre, I have adopted these in Table A—four in her central battery, two in barbette half-towers, and two on revolving platforms at the bow and stern respectively. She also carries eight light 5½-inch guns. This ship generally resembles her successors, the *Dévastation* and the *Foudroyant* (by the same designer), in so far as that her batteries fire past sides, with great tumble home.

Lord Brassey (in this respect somewhat erroneously following Mr. King, of the United States navy, in his able work upon *The Warships and Navies of the World*), says:

> The faculty of firing parallel to the line of keel is secured in the French ship by the tumble home of the ship's sides, and not by the projection of the battery beyond them, as in the English vessel (the *Audacious*).

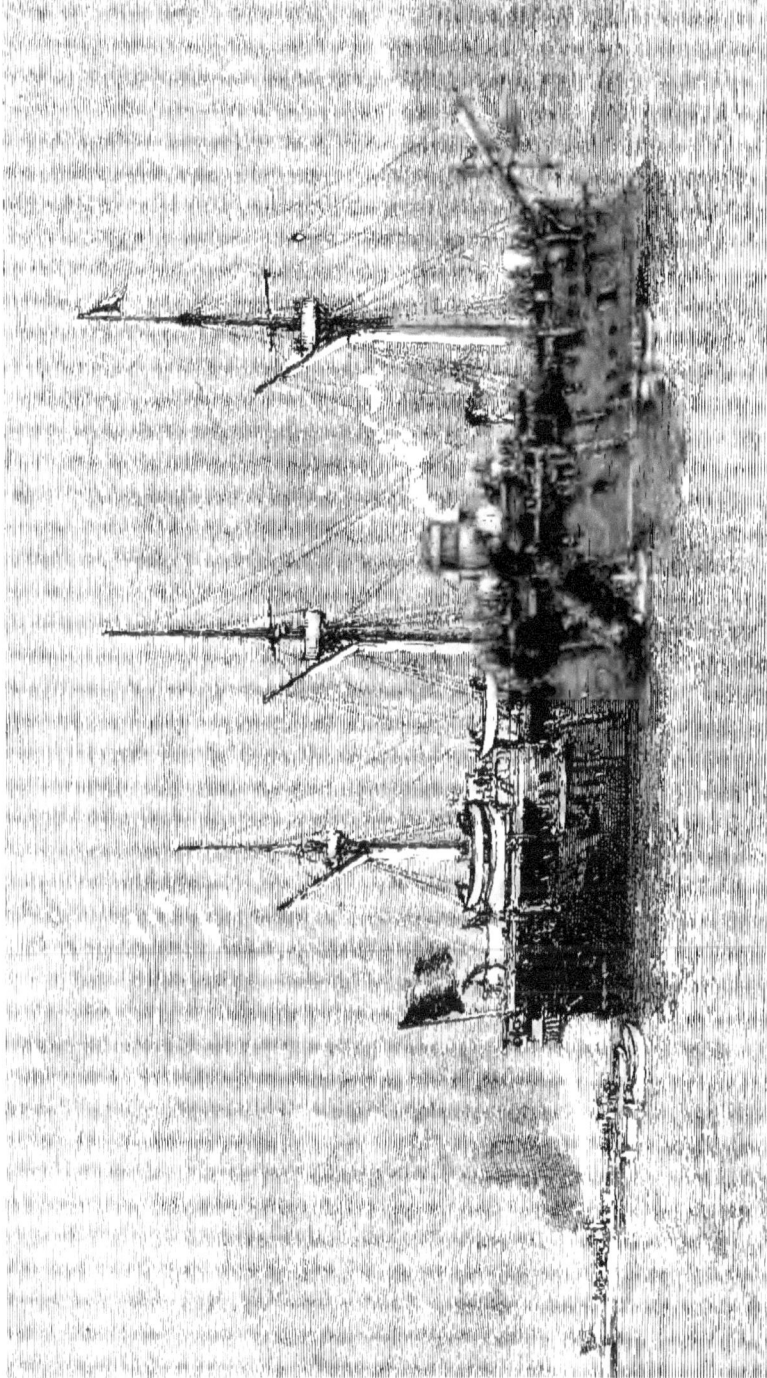

The Richelieu

It is difficult to understand what this means, because it is obviously only by the projection of the battery beyond the sides of the ship which are before and after it that fore and aft fire can be obtained from the battery in either case. But it is not true that the battery of the *Audacious*, any more than the battery of the *Redoutable*, projects beyond the breadth of the ship at the water-line, which would seem to be what is intended, and Lord Brassey may assure himself of the fact by looking at Plate III. of his own work on *The British Navy*, from which the above words are quoted. The *Redoutable* is a full-rigged ship, and nevertheless steams 14¼ knots per hour. There is one particular in which the *Dévastation* and the *Foudroyant*, like her as they are in general design, differ materially from the *Redoutable*. I refer to the armament. The former two ships each carry four 34-centimetre 48-ton guns in the main-deck battery, in lieu of the four 25-ton guns of the *Redoutable*.

The *Amiral Duperré* (designed by M. Sabattier, the able French chief constructor) claims a few words, as she differs materially in type from the three ships just discussed. She has a complete belt of very thick armour from stem to stern—greatest thickness 22 inches, tapering to 10 inches at the extremities, with a thick deck (2 inches) at the top of the belt in the usual manner. But above this belt there is no armoured main-deck battery, as in the other ships, the chief armament, of four 48-ton guns, being carried in four elevated barbette towers, two of which are well forward, and project considerably to enable their guns to act efficiently as bow-chasers, and at the same time to command all round the broadside and right astern. To facilitate this the sides of the ship have great tumble home. The other two towers are situated at the middle line of the ship, one near the stern, and the other farther forward, between the main and the mizzen masts. The main-deck, although without armour defence, is not without armament, as it carries fourteen 5½-inch 60-pounder rifled breech-loaders. Other particulars of the *Amiral Duperré* are given in the table, and in the following pages is a view of her, engraved from a photograph with which I have been favoured by a French officer.

It will be observed from her description that the most characteristic feature of this great ship of more than 10,000 tons is the absence of any guns protected by armour. The barbette towers, it is true, are armoured with 12-inch plates, and the main-deck guns are under the protection of the thin plating of the ship's side, which latter is of little or no avail, however, against the armament of other first-class ships. Practically the whole of the *Duperré*'s guns are unprotected. It may be added that during the discussions in London upon the "ships armoured in places" an attempt was made to show that the *Duperré*, owing to her alleged small initial stability, was as devoid of stability when injured above the belt as certain vessels of the British *Admiral* class when injured before and abaft the belt—a statement which I distrust, as I regard it as a mere inference from an experiment which I believe to be delusive. At the same time, the *Duperré* would have been the better for more initial stability.

But it is obvious that all belted or partially belted vessels, in which the belt is carried but a small height above the water for the size of the ship, must run the risk of losing both buoyancy and stability very soon if even moderately inclined in or after battle, seeing that, with a moderate inclination only, the entire armour-belt on the depressed side of the ship must disappear beneath the sea's surface. The strenuous assertion of this source of danger, although it could not lead to much increase in the stability of the existing armoured ships, has produced as one effect the busy and earnest efforts which both English and French constructors have been recently making to subdivide their ships *above the armour* into as many water-tight compartments as possible, and to stuff these compartments as full as possible of buoyant (or at least of water-excluding) materials. The necessity for resorting to this device, however, in first-class ships of nine, ten, or eleven thousand tons displacement, and of something approaching to five million dollars each in value, is not a thing for either French or English naval constructors to be proud of. But the assertion of the danger in question has had in England the further and very satisfactory result of bringing much more trustworthy ships, like the *Nile* and *Trafalgar*, into being, and of insuring the determined support of these ships in parliament whenever those who foolishly confound mere cheapness with merit in such constructions seek to interfere with the progress of these magnificent vessels.

Two other powerful ships of the French Navy, closely resembling the *Amiral Duperré*, are the *Amiral Baudin* and the *Formidable*. They are of 3¼ feet more beam than the *Duperré* (and therefore probably have much larger stability), and their displacement exceeds hers by 900 tons. Their armaments chiefly differ from hers in the employment of three guns of 75 tons each in their towers, in lieu of the four guns of 48 tons of the *Duperré*. The *Neptune, Hoche, Magenta,* and *Marceau* are four other powerful ships, as will have been seen from Table A, the principal armament of each consisting of four guns of 52 tons, carried in towers, with the exception of the *Hoche*, which has two of her four principal guns of 28 tons each only.

Incidental mention has already been made of two ships, the *Caï-*

THE ADMIRAL DUPERRÉ

man and *Indomptable*, which, although of only 7200 tons, carry very thick armour (19½ inches), and as a matter of fact carry also guns of the heaviest type (75-ton). There are two other vessels of the same description, the *Terrible* and *Requin*. Careful note should be taken of these four steel-built vessels, which add greatly to the power of France. Each carries two of the very powerful guns just mentioned, and steams at a speed of 14½-knots. In the same category of thickly armoured ships the French have yet one other ship, the *Furieux*, of 5560 tons. Her armour is 17½ inches thick in places, and she is armed with two 48-ton guns. Her speed is 12 knots. The *Tonnant* has the same armour and armament, but she is of nearly 1000 tons less displacement, drawing much less water, and steaming only at 10 knots per hour.

We may sum up the facts relating to the larger class of French ironclads which still rank among the efficient ships of 7000 tons and upward by saying that, in addition to the sixteen ships of which the particulars are given in Table A, there are on the efficient list the *Colbert, Friedland, Marengo, Ocean, Richelieu, Suffren, Trident, Savoie, Revanche, Surveillante,* and *Heroine,* most of which have been previously described in general terms, and the remainder of which are of less than 6000 tons, and were built chiefly of wood many years ago.

The French navy further comprises thirteen armour-plated cruisers, of which four have lately been dropped out of some official lists. Of the remaining nine, four are modern vessels, and all of about equal size and power. These are the *Duguesclin, Vauban, Bayard,* and *Turenne;* but of these, while the first two are built of steel, the last two are built of wood, with iron topsides, as are all the remaining five vessels of this class. The subjoined table will indicate the inferior character of most of the vessels of this type:

TABLE B.—FRENCH ARMORED CRUISERS.

Name of Ship.	Displacement.	Indicated Horse-power.	Speed.	Length.	Breadth.	Draught of Water.	Maximum Thickness of Armor.	Heaviest Guns carried.
	Tons.		Knots.	Feet.	Feet.	Feet.	Inches.	
Bayard.........	5900	4560	14.5	266	57.2	23.3	10	4 of 16 tons.
Duguesclin.....	5900	4000	14	266	57.2	23.3	10	4 " 16 "
Turenne........	5900	4250	14.2	266	57.2	23.3	10	4 " 16 "
Vauban.........	5900	4000	14	266	57.2	23.3	10	4 " 16 "
La Galissonière..	4700	2370	13	256	49	23	6	6 " 16 "
Triomphante....	4700	2400	12.8	256	49	23	6	6 " 16 "
Victorieuse.....	4600	2210	12.7	256	49	23	6	6 " 16 "
Reine Blanche ..	3620	1860	11.8	230	46.2	21.8	6	6 " 8 "
Thetis..........	3620	1860	12	230	46.2	21.8	6	6 " 8 "

Of the above ships it may be remarked that the *Thetis* and *Reine Blanche* have been nearly twenty years afloat, the *Galissonière* was launched in 1872, the *Victorieuse* in 1875, and the *Triomphante* in 1877. The remainder of the nine, as previously stated, are modern vessels, the *Duguesclin* being not yet completed. The *Duguesclin* and her sister ships are of the *Duperré* type, much reduced in dimensions.

There are nine completed coast-guard ironclads and eight armoured gun-boats in the French navy, as follows:

TABLE C.—FRENCH IRON-CLAD COAST-GUARD VESSELS.

Name of Ship.	Displacement.	Speed.	Maximum Armor.	Principal Guns.	
	Tons.	Knots.	Inches.	No.	Tons.
Fulminant	5600	13.22	13	2	28
Tonnerre	5700	14	13	2	28
Tempête	4523	12	13	2	28
Vengeur	4523	10.8	13	2	48
Bélier	3600	12.3	8.5	2	16
Bouledogue	3800	12.25	8.5	2	16
Cerbère	3800	11.4	8.5	2	16
Taureau	2700	13	6	1	23
Tigre	3500	13.5	8.5	2	16

TABLE D.—FRENCH IRON-CLAD GUN-BOATS.

	Name of Ship.	Displacement.	Speed.	Maximum Armor.	Principal Guns.	
		Tons.	Knots.	Inches.	No.	Tons.
First Class.	Achéron	1639	13	8	1	28
	Cocyte	1639	13	8	1	28
	Phlégéton	1639	13	8	1	28
	Styx	1639	13	8	1	28
Second Class.	Flamme	1045	13	8	1	16
	Fusée	1045	13	8	1	16
	Mitraille	1045	13	8	1	16
	Grenade	1045	13	8	1	16

The vessels in the tables C and D are all revolving turret vessels, with the exception of the *Taureau* and of the four second-class gun-boats, which fire their guns *en barbette*. They embrace very different types of construction, involving different degrees of sea-worthiness—very low degrees in some of them, I fear. With the exception of the *Tempête*, they are all furnished with twin screws. The *Fulminant, Tonnerre, Tempête* and *Vengeur*, in Table C, and the whole of the vessels in Table D (as yet incomplete), are of iron or of steel, or of the two combined; the remainder have hulls principally built of wood. I have chosen for illustration the turret-vessel *Vengeur*, as seen in the following pages, which has been engraved from a photograph sent to me by a naval friend in France.

We come now to the unarmoured ships of France, and as in writing of these I purpose accepting the official classifications adopted in France, which are not identical with those employed in England, it may be well to repeat here a caution which the British Admiralty has given in a memorandum prefixed to a recent "return" of theirs "showing the fleets of England, France, Russia, Germany, Italy, Austria, and Greece." The caution is to the effect that France includes under the heading of "cruisers" vessels of about similar value to the larger class of English sloops, which are excluded from the English "cruiser" class. But I regret the necessity of observing that the Admiralty officers, while careful to put this explanation well forward, appear to be equally careful to withhold an explanation of much greater moment concerning three French cruisers of large size and of greater importance—withheld in pursuance, apparently, and as I have most reluctantly come to fear, of an uncandid, and indeed of a misleading spirit, which seems to have taken possession of some persons who have to do with the preparation of Admiralty returns to parliament. The exercise of this spirit has forced me ere now to draw the attention of parliament to the matter, and in one instance to have an official return, which contained erroneous and too favourable classifications of British ships, withdrawn.

Anyone referring to the parliamentary return of British and foreign fleets just adverted to will find under the heading of "unarmoured vessels building" two large and remarkably fast steel cruisers, the *Tage* and the *Cecile*, the former of which exceeds 7000 tons in displacement, while the latter approaches 6000 tons, and both of which are to steam at the immense speed of 19 knots an hour, or a knot in excess of the fastest armed vessel (neglecting torpedo craft) in the British navy. These two French cruisers are respectively 390 and 380 feet in length, and are to be driven by over 10,000 indicated horse-power in the *Tage*, and by nearly 10,000 indicated horse-power in the *Cecile*. A third vessel, the *Sfax*, launched at Brest in 1884, of 4420 tons, 7500 indicated horse-power, and 16½ knots speed, is also given without remark in the parliamentary return as an "unarmoured" vessel. Now even this last-named vessel has a steel deck one and three-fourths inches thick to protect her boilers, machinery, and magazines, while the *Tage* and *Cecile* have such decks three inches thick. These, being mere decks, do not, of course, remove the ships out of the category of unarmoured ships, and the return is correct in this respect. But now in this same return all the British ships provided with protecting decks of this character are kept out of the lists of unarmoured or "unprotected" vessels, and are classed separately, and are described as "protected" vessels. And not only is this true of vessels like the *Mersey* class, which have such decks two and one-half inches thick in places, but it is true likewise of some twenty vessels, ranging, many of them, as low as 1420 tons in displacement, and with decks and partial decks of less thickness than that of the *Sfax*, the weakest of the three French ships in this respect.

In short, while the twenty-two English ships are withheld from the category of unarmoured ships, although every one of them is inferior in protecting decks to the three French ships, the latter are placed in the inferior category, and not a word of explanation is offered to prevent the uninitiated and unsuspecting reader from regarding as weaker than our vessels those French vessels which are in fact the strongest and best protected. I must say that, as an Englishman, I grieve to see returns to the British parliament made use of for the dissemination of information so misleading as this; and I should do so if I could believe there was nothing but official negligence involved; but I am sorry to say I cannot doubt that had the mere reproduction of foreign classifications put three of the very fastest and most important cruisers of our own navy, of Admiralty origin, at the very great disadvantage to which the French ships are put in this return, we should have had a very full and a very prominent explanation of the seeming discrepancy given. It is to the credit of *Lloyd's Register* office that what the Admiralty Office failed to do in a paper issued at the end of July was properly done in their *Universal Register*, published two or three months earlier; for in the latter the three French ships are separately detailed under the heading of "deck-protected cruisers."

It is absolutely necessary to bring to light the matter just explained, for otherwise the present state and the prospects of the French navy cannot be properly understood, the *Tage*, *Cecile*, and *Sfax* being, on the whole, the most important of the French ships which are without armour belts. Two others there are, however, which are weaker than the *Tage* and *Sfax* only in the fact of their being without special deck protection. These are the *Duquesne* and the *Tourville*, two ships approximately alike in size and construction, and both having their iron bottoms sheathed with two thicknesses of wood and then coppered, after the manner introduced by myself in H. M. S. *Inconstant*. Both of these French ships have attained $16^{9/10}$ knots of speed. They are armed with seven guns of eight tons and fourteen of three tons weight.

The remaining unarmoured vessels of France must be rapidly summarized. It is impossible to neglect in this case, as was done in my article on the British navy, all the frigates, etc., which have frames of timber, because to do this would be to omit all unarmoured frigates of the French navy except the *Duquesne* and the *Tourville*, already described. But it is not necessary to do more than name the *Venus, Min-*

erve, and *Flora*, all launched prior to 1870, and all slow, and to say that there remain but four unarmoured wood frigates of 14 knots speed, of about 3400 tons, and armed with from two to four guns of five tons, and eighteen to twenty-two guns of three tons. These are *Aréthuse, Dubourdieu, Iphigénie,* and *Naïade*, which, although wooden ships, have all been launched since 1881—the *Dubourdieu* in 1884. Of French first-class cruisers which do not rank as frigates (having no main-deck batteries) there are nine in number, all built of wood except one, the *Duguay-Trouin*, which is the fastest of them all, steaming at $15\ ^9/_{10}$ knots. This vessel has 3300 tons displacement, and is armed with five guns of eight tons and five of three tons. None of the remaining eight exceed 2400 tons in displacement, none exceed 15.3 knots in speed (but none are less than 14 knots), and each of them carries fifteen guns of three tons. Next come thirteen second-class cruisers, ranging in displacement between 1540 and 2100 tons, and in speed between 11½ and 15 knots; they are principally armed with 3-ton guns. There is another vessel, the *Rapide*, in this class, but I only know of her that her tonnage is 1900 tons. Of cruisers of the third class there are fifteen, ranging from 1000 to 1400 tons, and principally armed with 3-ton guns.

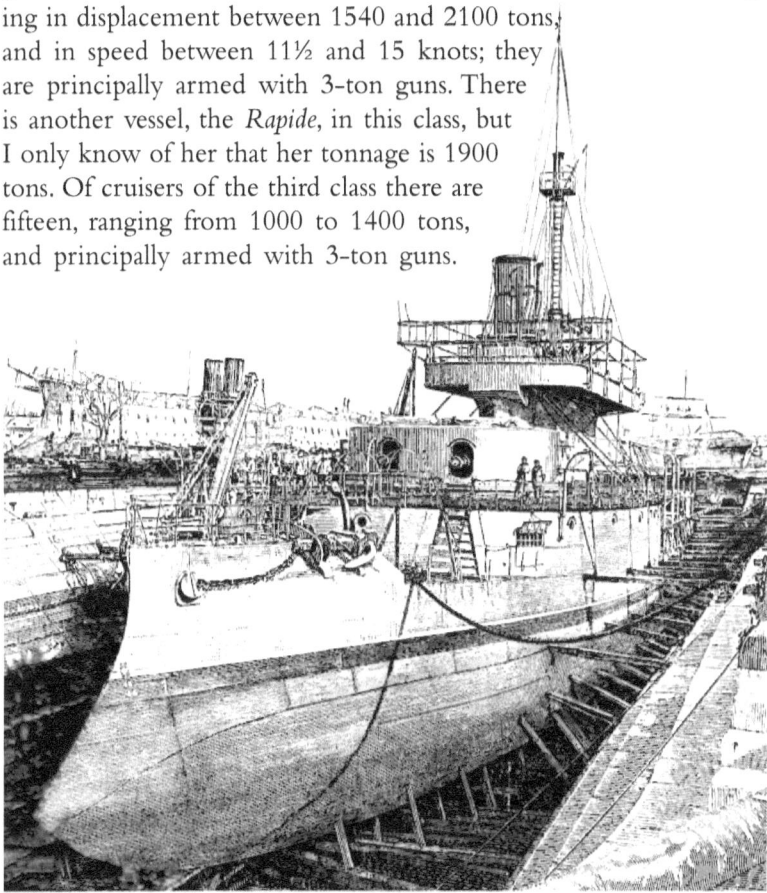

THE VENGEUR: FRENCH IRONCLAD COAST-GUARD VESSEL

Their speeds vary from 10 to 13 knots; one, however, the *Hirondelle*, steaming at 15½ knots. The French have likewise thirty-five vessels, "*avisos*," etc., of which about one-half are from 1400 to 1600 tons, and the remainder are from 720 to 1000 tons. About six of them reach or approach 13 knots, but most of them range between 10 and 11 knots, some of them falling as low as 8 knots. I have further to make mention of two very fast vessels—for they are to steam 19½ knots—now under construction, named the *Surcoup* and the *Forbin*, each of 1850 tons, and each armed with two 3-ton guns. There is also a vessel of 1540 tons, named the *Milan*, which steams 18 knots, and is armed with five very light (24-cwt.) guns. The French navy possesses also ninety-nine vessels, most of them carrying guns (many of 3 tons, some of 5 tons, and one or two of 8 tons), and also twenty-eight steam transports, varying in size from 1200 to nearly 600 tons, the largest of them, the *Nive* (of 5680 tons), steaming 14 knots.

The navies of Europe, including the British navy, have undergone of late considerable expansion in respect of their very fast unarmoured steel vessels, the designing and successful construction of which have been brought about by improvements in the quality of ship steel and in steam machinery, notably as regards the latter, by the employment of "forced draught." These are called torpedo-vessels, as distinct from torpedo-boats. There are in process of completion for the British navy eight of 1630 tons (the *Archer* class), each carrying six 6-inch 5-ton guns, and estimated to steam with forced draught from 16 to 17 knots; two of 1430 tons each (*Scout* class), carrying four 5-inch 2-ton guns, with an estimated maximum speed of 16 knots; and two of 785 tons (*Curlew* class), called "gun and torpedo" vessels; speed, 15 knots; armament, one 6-inch 89-cwt. and three 5-inch 36-cwt. guns. There is also a class of "torpedo gun-boats "(the official designation, but not one which expresses any very manifest distinction from the last-named class), which are of a very notable character. This (the *Grasshopper*) class, of which each vessel is of only 450 tons displacement, is to be supplied with engines of 2700 indicated horse-power. The diagrams following exhibit the general form and particulars of these very remarkable little vessels, which are expected to steam at fully 19 knots (22 miles) per hour. Against the above torpedo-vessels of the British navy are to be set, in the French navy, four torpedo cruisers of 1280 tons, 17 knots speed, carrying each five 4-inch guns; and eight torpedo despatch vessels, each of 320 tons, and designed to steam at 18 knots, carrying machine guns only; such machine guns being also carried, of course, by all the fast

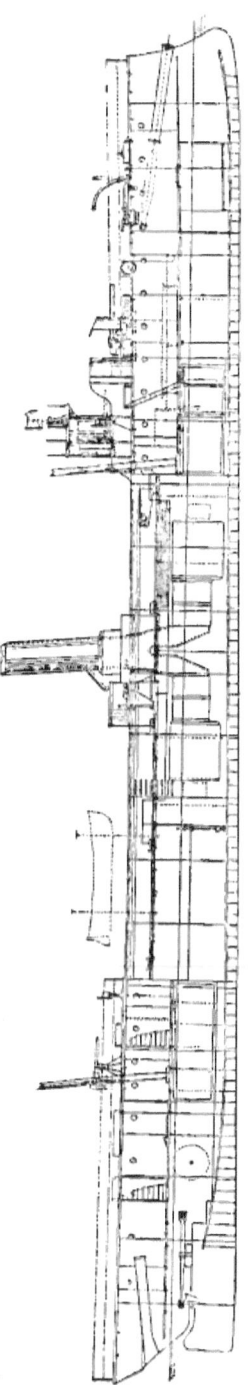

British torpedo boat of the Grasshopper class (side view)

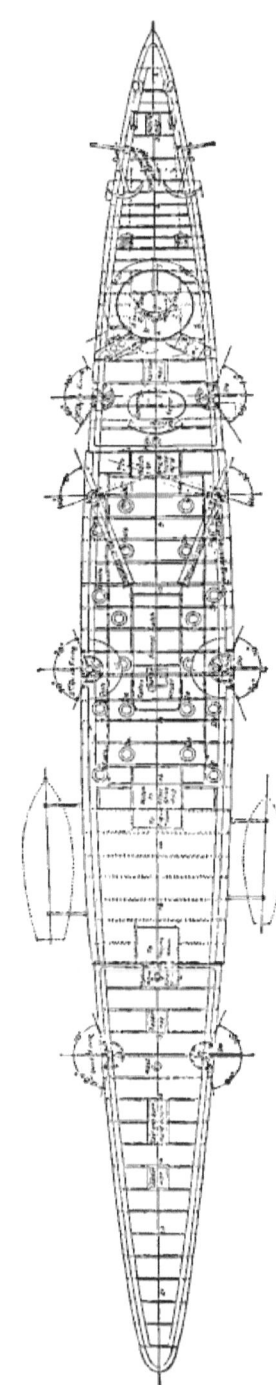

British torpedo boat of the Grasshopper class (upper deck, poop and forecastle)

torpedo-vessels and gun-boats, both French and English, previously referred to, but in their cases in conjunction with their other guns. These 320-ton torpedo-vessels of France are to be driven by machinery of 1800 indicated horse-power.

It may be observed with regard to these small craft furnished with such enormous steam-power (in proportion to their size and tonnage) that there is much uncertainty as to the speeds which they will attain. Not only are the builders without experience of similar vessels by which to guide themselves, but where the proportion of power to displacement is so great, slight differences both in hulls and machinery, no less than in immersion and trim, may produce unforeseen results. As designers who fail to realize promised speeds are liable to be discredited, while those whose vessels surpass their promised speeds may be unduly praised, it is but reasonable to expect that the promised speeds will usually even be more than realized. This has been the case with the *Bombe*, the first of the French torpedo despatch vessels which have been tried under steam, and which, under the promise of 18 knots, realized no less than 19½ knots on the measured mile. It should be added that all of these extremely fast small craft in both navies are propelled by twin engines and screws. As great public interest will be felt in the trials of these very novel and special vessels—as mere steamers no less than as war craft—it may be well to give their names, to facilitate their identification hereafter.

English Torpedo Gun-boats: *Grasshopper, Rattlesnake, Spider, Sandfly*—each having a displacement of 450 tons, 2700 horse-power, 200 feet length, 23 feet breadth, 8 feet draught, and a speed estimated at 19 knots.

French Torpedo Despatch Vessels: *Bombe, Couleuvrine, Dague, Dragonne, Flèche, Lance, Saint-Barbe, Salve*—each having a displacement of 320 tons, 1800 horse-power, 194.3 feet length, 21.4 feet breadth, 5.1 feet draught, and, with the exception of the *Bombe*, a speed estimated at 18 knots. The actual speed of the *Bombe* is 19.5 knots.

Besides the above vessels, the two navies (English and French) are provided as follows with torpedo-boats: The English have nine small (56 feet long) and slow (14½- to 15 knots) of wood; fifty small (60 to 66 feet long) and slow (15 to 16 knots) of steel; nineteen others of greater length, but all less than 93 feet, and of speeds varying from 16 to 19 knots; six of 100 to 113 feet, and 19 knots; fifty-three of 125 feet in length, and 19 knots; and two building, *viz.*, one of 135 feet in length, and 22 knots, and one of 150 feet in length, and 20 knots; in all, one hundred and thirty-nine torpedo-boats, of which the 135-

feet boat carries four 3-pounder quick-firing guns, and the 150-feet boat carries five 6-pounder guns of that land. The French have nine under 70 feet in length; forty-one under 100 feet in length, steaming at 17 to 18 knots; eighteen of 108 feet in length, somewhat faster; nine of 113 feet in length, steaming at 22 knots; and fifty-one of 114 feet in length, steaming at 20 knots; in all, one hundred and twenty-eight torpedo-boats, all armed with machine guns only. As the nine slow wooden boats of the English navy can hardly be regarded as torpedo-boats at all, it may be said that of torpedo-boats, built and building, the English have one hundred and thirty, and the French one hundred and twenty-eight, of which the English have seventy-nine completed, and fifty-one building and completing, and the French have sixty-eight completed, and sixty building and completing. The English navy is therefore slightly, but only slightly, in advance of the French in the matter of torpedo-boats proper, while in respect of extremely fast sea-going torpedo-vessels of 320 and 450 tons respectively, the English have three under construction and one completed, while the French have one (the *Bombe*) completed and seven under construction.

Notes

Of the 150,000,000 *francs* appropriated in France this year for the construction of warships nearly nine-tenths were set aside, not for the building of large armoured vessels, but for the following fast cruisers and auxiliary classes:

Six cruisers, class 1, 30,000,000 *francs*; ten cruisers, class 2, 26,000,000 *francs*; twenty torpedo-catchers, 12,-000,000 *francs*; fifty gun-boats, 15,000,000 *francs*; one hundred torpedo-boats, 25,000,000 *francs*; three coast-defence vessels, 25,000,000 *francs*.

Notwithstanding the late change in administration this seems to show that the policy of Admiral Aube, referred to in the introductory chapter, is still potent, and that the government believes the next war with England will be carried on by French cruisers attacking British commerce, and that sharp, destructive dashes will be made against the enemy's coast by ships with great speed, and such sufficient power that "all of England's littoral towns, fortified and unfortified, whether purely peace establishments or warlike," will be burned or pitilessly ransomed. This exponent of the new ideas continues:

In any future war France will come down from the heights of the cloudy sentimentality which has created that monstrous association of words, *rights* of war, and her attack on every source of English riches will become not only legitimate but obligatory.

It is certain that French naval activity is now mainly directed to the construction of vessels just suited to these new theories. At the same time she has a formidable fleet of heavily armoured vessels, a rough comparison with those of England being as follows, in the classes which have over fifteen inches of armour protection and carry guns above forty-three tons in weight:

Ships.	Armor.	No. Guns.	Weight.
3	21¼ inches	6 breech-loading rifles	75 tons.
5	20 "	8 " "	75 "
6	18 "	8 " "	50 "
2	15 "	24 " "	48 "
16		46	

England has three ships with armour from twenty-four inches to eighteen inches thick, and twelve ships with armour eighteen inches thick; three ships carry six 110-ton guns, six carry twenty-four 67-ton guns, five carry sixteen 43-ton guns, all breech-loading rifles, while one has four muzzle-loading 80-ton guns.

The latest additions to the armoured fleet of France are the *Hoche*, of the Marceau type of battleship, and the *Cocyte* and *Mitraille*, coast-defence gun-boats. The *Marceau*, launched May 24, 1887, is built of steel with an under-water skin of iron; a double bottom extends below the engines, boilers, and magazines, and the hold is divided into thirty-one water-tight compartments by horizontal and longitudinal bulkheads. The armour-belt encircles the ship, dips forward to strengthen the ram, is carried twelve inches above the load water-line, and varies in thickness from 13.7 to 17.7 inches; the barbette towers are 15.7 inches thick, and the armoured deck, above which there are many compartments, is 2.6 inches thick. The armament is made up of four $13^{4/10}$-inch guns mounted in the towers, of one 5½-inch gun at the bow, and of sixteen 5½-inch pieces in broadside; the secondary battery includes twenty Hotchkiss guns and four above-water torpedo tubes.

The estimated horse-power is 8548 (not 5500 as stated previously) with natural draft, and 12,000 with forced draft, the estimated speed being 16 knots, the coal capacity 800 tons, and the coal endurance 1500 miles at full power and 3500 miles at 11 knots speed. The *Neptune* and the *Hoche*, of the same general plans and dimensions, were launched in the spring of this year. As originally designed the *Hoche* was expected to develop 16 knots and 7000 indicated horse-power, but by the application of forced draft the speed was increased to 17½-and the power to 12,000. The armament consists—not of the four 52-ton guns given in the table on page 76—but, as stated in the text, of two 13.4-inch guns (34 centimetre) mounted one in each of the midship turrets, of two 10.6-inch guns (27 centimetre) carried one in each of the waist turrets, and of eighteen 5.5-inch guns (14 centimetre) so disposed in broadside within the unarmoured central superstructure which occupies the deck between the turrets that the forward and after pairs are given bow and stern fire respectively. The armour-belt is similar to that of the *Marceau*, but the protective deck is from 3.15 to 3.54 inches thick, and the heavy gun sites are protected by 15.75 inches of compound armour.

The *Amiral Courbet* (formerly the *Foudroyant*) carries four 10.6-inch, six 5.5-inch, and twelve rapid-fire guns. She developed 6016 horse-power with natural and 8088 with forced draft, the mean speed being 14.2 knots on a consumption of 2.35 pounds per power each hour. The *Indomptable, Requin, Caïman,* and *Terrible* are sister battleships. They were originally laid down in 1877, and the *Terrible* was only completed ready for sea in 1887. They are constructed like the *Marceau*, of iron and steel, the outer skin of the under-water body being of the former metal; the compound armour is from 13 to $19^{5/8}$ inches in thickness, and carries five feet of its seven feet six inches width below the water-line. In each of two pear-shaped barbette towers situated on the longitudinal midship line, and protected by 17¾ inches of armour, a 16.5-inch gun, with its axis twenty-one feet above the water is mounted; in addition there are four 4-inch breech-loading rifles and a secondary battery of rapid-fire guns and torpedo-tubes. The *Indomptable*, launched in September, 1883, made in her trial trip in August, 1886, a speed of 15 knots, and is officially rated as having a sea speed of 13.5 knots. All work upon the partially protected ships *Brennus* and *Charles Martel* was stopped in 1886, and their specific appropriation has been transferred to the sum already assigned for the construction of fast cruisers and torpedo-boats.

The *Cocyte* and *Mitraille* belong to a new class, or rather they represent a type which, after disappearing for a season of doubt and denial, has had its value so much recognized that three Continental nations are giving it earnest study. A late French minister of marine asked within a year for money to construct fifty of these gun-boats, but was then refused the grant, a decision for which Admiral Sir George

Elliot thinks England ought to be very grateful. This distinguished officer believes in the value of the type, and hopes that the Admiralty "will take note of the threat thus made" before the theory is allowed to prevail that adequate security can be given to the British coasts by sea-going cruisers, submarine mines, shore batteries, and torpedo-boats. The boats present a small target, and give good armour protection to guns which, when the vessels are inshore or reinforced by land batteries, have sufficient power to keep battleships at a distance. They are very handy, have good speed, and are economical, because for the same money they can, as flotillas, bring into the action four times the gun power possible in the large battleships.

In France this type is divided into two classes—the *Acheron, Cocyte, Phlegéton,* and *Styx,* of 1639 tons, belonging to the first, and the *Fusée, Grenade, Mitraille,* and *Flamme,* of 1045 tons, to the second. The iron and steel hulls are extensively subdivided into water-tight compartments, and are protected by complete belts of steel armour at the water-line, and by arched steel-armoured decks. The superstructures above the protective decks have water-fine belts of cellulose. The armament consists of one heavy gun mounted in a barbette tower, and of a strong secondary battery of machine guns and torpedoes.

The most important contributions to the sea-going navy of France are the cruisers. In the naval programme adopted after the war with Germany, ships of high speed were decided to be of such great value that thirty-four—sixteen of the first and eighteen of the second class—were provided for. At the present day French naval policy seems to pin its faith to fast cruisers, 5½-inch breech-loading guns, and torpedo-vessels. In pursuance of this belief the *Tage,* the largest unarmoured cruiser yet designed by any nation, was laid down in 1885; she is ship-rigged, has a complete underwater curved deck, lightly armoured bulk-heads forward and abaft the battery, a steel conning tower, and heavily plated hatchways. A belt of cellulose along the water-line, and the sub-division of the space above the protective deck into water-tight compartments, will, it is claimed, insure the safety of the ship in action. This employment of cellulose to stop leaks automatically was very successfully demonstrated in the experiments made at Toulon with a target:

> Composed of fourteen parts of cellulose and one part of cellulose in fibre, the whole compressed into a felt-like mass, with a lining two feet thick. A shot seven and one-half inches in calibre was fired against this target at a distance to insure

penetration. The result was not only satisfactory but extraordinary. The shot, which carried away about one-fifth of a cubic foot of the composition, had no sooner passed through than the cellulose closed up so firmly that a strong man was unable to insert his arm into the hole. A tank filled with water was then hung against the place where the shot had entered, and after an interval of fifteen minutes water began to trickle through, but not more than a man with a bucket could easily intercept. As soon as the composition became thoroughly soaked, it offered increased resistance to the entrance of the water, which eventually ceased to flow, and the breach was closed automatically. The results were the same where shells were used in-stead of shot, and red-hot coals were heaped upon the composition with-out causing its ignition. (Lieutenant Chambers, U. S. Navy.)

The twin-screw cruiser *Cecile*, which was designed before the *Tage*, and is somewhat smaller, illustrates the principle of duality in construction; the two main engines are situated in separate compartments, and the six boilers are arranged in three different groups. The sail area is 2153 square yards, and the steel lower masts serve as ventilators to the hold, and carry steel crow's-nests in which are mounted rapid fire and machine guns. The primary batteries of the two ships are similar, each carrying six 6¼-inch guns on the spar deck (one forward, one aft, and four on sponsons) and ten 5½-inch pieces on the covered deck in broadside. The secondary battery of the *Cecile* consists of ten 37-millimetre (1.45-inch) guns, and that of the *Tage* of three 47-millimetre (1.85-inch) rapid-fire guns, and twelve 37-millimetre revolving cannon—all of the Hotchkiss pattern. Both ships are supplied with above-water torpedo tubes, the former having four, the latter seven. The estimated maximum speed of the *Tage* is 19 knots, with 10,330 horse-power, and that of the *Cecile* is 18½ knots, with 9600 horse-power. The latest cruisers laid down are the *Jean Bart* and the *Dupuy de Lôme*, the first bearing the name of the rugged old sea-wolf who entered the navy as an apprentice and died a famous admiral, and the other that of the constructor who designed both in wood and iron the first steam line-of-battle ships. These vessels are of 352 feet length, 43.6 feet beam, 18 feet 10 inches mean draught, and 4162 tons displacement; their estimated maximum speed is 19 knots. The main battery is composed of four 6.3-inch guns mounted on sponsons, and of six

5.5-inch carried in broad-side, and the secondary armament has six 37-millimetre revolving cannons, four 3-pounder rapid-fire guns, and the usual torpedo tubes.

The *Alger* and *Isly* are similar in construction to the *Cecile*, but have the dimensions and armament of the *Jean Bart;* they are designed for 19 knots, and a coal endurance of 3600 miles at 13 knots. The *Mogador* is a rapid cruiser of 4325 tons, and of nearly similar design, armament, speed, and endurance as the above. The *Chanzy, Davoust,* and *Suchet* belong to the same class of "*croiseurs à barbette*," and are of 3027 tons displacement, with an estimated speed of 20 knots.

The *Surcoup* and *Forbin* illustrate another favourite type of cruiser. They are 311 feet 7 inches long, have 30 feet 6 inches beam, and on a mean draught of 13 feet 11 inches displace 1848 tons. The hull weighs 817 tons, and the engines (with boilers filled) 544 tons; the coal capacity is 200 tons, and the endurance 2400 miles at 10 knots. The engines are expected, with forced draft, to develop 6000 indicated horse-power and 19.5 knots. They have a four-masted schooner rig, spread 7255.5 square feet of canvas, and carry a complement of one hundred and fifty officers and men. The battery consists of two 5.5-inch guns on the upper deck, three 47-millimetre rapid-fire guns on the poop and forecastle, four 37-millimetre Hotchkiss revolving cannon on the rail, and five torpedo launching tubes—two firing ahead, one astern, and one on each beam. This lightness of battery and small coal capacity indicate with great precision how much weight-carrying power has been sacrificed to spars and sails. The *Coetlogon* and *Cosmao* laid down this year are of the same type.

Wishing to obtain a small class of steel cruisers, the French Government lately invited the leading ship-builders to send in competitive designs for a vessel which at an extreme draught aft of fourteen feet would on the least possible displacement sustain with natural draft a speed of eighteen knots for twelve hours, and with forced draft a speed of nineteen knots for two hours. The coal endurance was to be 2400 miles at ten knots, the main battery to include two 5.5-inch guns, and the protective steel deck to be 1.6 inch thick. Five competitors furnished plans, and finally those of *the Societé de la Gironde* were chosen, and the two vessels now known as the *Troude* and the *Lalande* were laid down. Their principal dimensions are, length 311 feet 7 inches, beam 31 feet, mean draught 14 feet, and displacement 1877 tons. The armament will be two 5.5-inch and three rapid-fire guns, four 37-millimetre revolving cannon, and a supply of torpedo tubes. The vessels, as with the *Surcoup* type to which they are very similar, will have a fore and aft rig and a complement of one hundred and sixty.

The *Duguesclin*, referred to earlier, is an armoured cruiser built of steel and iron and sheathed with wood and coppered; an iron armour belt $9^{1/8}$ to 6½ inches thick encircles her, and the four barbettes are protected by 8 inches of compound armour. The armament consists of four 9.45-inch guns in the barbettes, of one 7.5-inch gun in the bow, and of six 5.5-inch pieces on broadside, in addition to two 6-pounder rapid-fire guns, twelve revolving cannon, and two above-water torpedo tubes. Her displacement is 5869 tons, draught 23 feet 3 inches, and she has developed 4100 horse-power and 14 knots. The *Sfax* is a partially protected steel cruiser, which is sheathed with wood and coppered, and has an underwater curved steel protective deck 1.5 inches thick. There is the usual water-tight subdivision below and above this deck, together with the lately adopted cellulose belt. The armament consists of six 6.3-inch guns, four mounted on sponsons, with bow and stern fire, and two in recessed ports with bow and beam fire; of ten 5.5-inch guns in broadside on the main deck, and of eight Hotchkiss revolving cannon. In May, 1887:

> With natural draft, the mean indicated horse-power developed for four hours was 4333, and the speed 15.9 knots. With forced draft the mean results of a six hours' trial were, indicated horse-power, 6034; revolutions, 78; mean speed, 16.84 knots. The trials for coal endurance showed that with full natural draft speed the consumption was 1.96 pounds per hour per indicated horse-power developed, and with forced draft it was 2.10 pounds. During these trials the draught of water was 19 feet 4 inches forward and 25 feet 1 inch aft, which was in excess of the normal designed draughts of 19 feet 8 inches and 24 feet 8 inches. Notwithstanding this fact, and the fact of the indicated horse-power falling much below the estimated power of the engines (5000 with natural, 7500 with forced draft), the speed realized exceeded the maximum estimated of 16.5 knots. (Lieutenant Colwell, U. S. Navy.)

France has been very active in the construction of torpedo-vessels. On the present plane to which the science of naval warfare has advanced the great tactical question is whether torpedo-boats or flotillas are, in high-sea duels or engagements, to take the place of huge ships or large fleets. There are, even in France, very marked differences of opinion upon this point, but so far as official policy and programme can assert a belief, there is no other nation, Russia alone excepted, which appears to hold the torpedo in such-high esteem. The manoeuvres of 1886 were notable for the prominence given to that type, and of the forty vessels assembled this year for drill and instruction at Toulon, twenty-one were torpedo craft of some kind. The French navy has over two hundred torpedo-boats, which vary in length from seventy to one hundred and thirty-three feet, and in speed from fifteen to twenty-three knots; England has one hundred and eighty-one, of which eighty-eight are built and ninety-three are under construction; these differ as much among them-selves as the French boats, their speed range being about the same, and their lengths varying from sixty-three to one hundred and fifty feet. Generally described torpedo-boats may be divided into two classes, the first including such as are of a size to keep the sea and act independently, and the second those carried by ships. The Whitehead torpedoes, the type most generally used, are ejected from their firing tubes by various means, slow burning power being employed in some cases, though more frequently compressed air or steam. The success of the French boats in China has revived the use of the spar torpedo in combination with the locomotive type, and with us the promised success of the Howell design may cause another revolution in this system of attack. White writes:

> Boats exceeding one hundred feet in length have been shown capable of making long sea voyages unaccompanied, and the fact has been seized upon by enthusiasts in torpedo warfare like the late Gabriel Charmes as evidence that the days of the armoured ship, of the large and costly cruiser, were numbered. Actual experience is not favourable to this extreme view. There is a clear and marked distinction between the capability of making long sea passages in safety, when specially prepared for the purpose, and the sure sea-going qualities of large ships. Boats of the largest size and small swift vessels cannot equal large vessels in the power of maintaining their speed and fighting efficiency or rough war. Life is scarcely endurable for long periods in these overturning boats and small craft, cooking is often a difficulty, and it is not every officer who can rival the foreign commander of a torpedo-boat I once met, who had acquired the power of living for long periods on sherry and eggs. M. Weyl stated a fact when he said of the grand manoeuvres with the French ironclads and torpedo flotilla last year (1886), 'In my experience as a sailor I have always found that the sea is

merciful to big ships and hard upon small ones.' A moderately rough sea that scarcely troubles the ironclad or the cruiser of consider-able size, suffices to render inevitable a reduction in speed of the small vessels, and a serious loss of power in the accurate use of their torpedoes and guns. As adjuncts to fleets, the small swift vessels and boats are undoubtedly of immense value under many circumstances; for the defence or attack of forts and coasts they are well fitted, but as substitutes for all other types, and as the successful rival of large warships in sea service, their claim is not, and probably will not be, established.

The discovery of the minimum size of swift torpedo-vessels or torpedo-boat destroyers really capable of independent sea service with a fleet is now engaging attention in all navies. In France the first attempts were made in the *Bombe* class in 1883; some vessels of this class were tried in the recent manoeuvres and favourably mentioned. In England the *Grasshopper* class was designated in 1885, and the first completed vessel, the *Rattlesnake*, is now completing her speed trials.

Since this last sentence was written the *Rattlesnake* has made 18.799 knots with a collective indicated horse-power of 2718.27, and though the weather was boisterous, proved that under normal conditions she could furnish a fairly steady platform for her battery. Chief constructor White continues as follows:

These vessels are of 450 tons, and estimated to steam about 19 knots an hour. Messrs. Thomson, of Clydebank, have just completed another example of the class, intermediate in size between the *Bombe* and *Grasshopper*, and said to have attained the very high speed of 22½ knots on trial in smooth water. Experience at sea with these vessels will be of immense value to future designs. They combine an armament of light guns with torpedo armaments, and can act either as torpedo-vessels or as destroyers of torpedo craft. Similarly in the largest classes of torpedo-boats light guns as well as torpedoes are provided for. In fact there has been a departure from the original idea of having the torpedo as the only weapon, as the boats have increased in size, and this change cannot but commend itself.

The *Milan* mentioned in the text was designed in 1879 for a torpedo despatch-vessel, but is now used as a scout. She is 303 feet in length, 32 feet 10 inches in beam, draws 15 feet 1 inch aft, and has a displacement of 1550 tons. She carries a fair battery but no torpedoes, is propelled by twin screws, each worked by two compound tandem engines, and has Belleville boilers. On her trial she made 18.4 knots in a rough sea, and developed with natural draft 4132 horse-power, or more than was expected with forced draft; she carries three hundred tons of coal and has a three-masted schooner rig.

The rapid development of torpedo-vessels since her day has resulted in the evolution of different types suited to different demands, and of late France has adopted the following classification for her torpedo flotilla:

		Displacement.
1.	Torpedo cruisers (*croiseurs-torpilleurs*)	1260 to 1280 tons.
2.	Torpedo despatch-boats (*avisos-torpilleurs*)	320 to 360 tons.
3.	Sea-going torpedo-boats (*torpilleurs de haute mer*)	50 tons and over.
4.	Coast-guard torpedo-boats (*torpilleurs-garde cotes*)	*a*, 25 tons; *b*, 50 tons.
5.	Picket torpedo-boats (*torpilleurs-videttes*)	less than 25 tons.

The *Condor, Epervier, Faucon,* and *Vautour* are examples of the first class, and combine the lightness of hull and the gun armament of the torpedo-catcher with the sea-going powers of the cruiser. They are twin-screw steel vessels, 216 feet long, 29 feet 2 inches in beam, 15 feet 5 inches in draught, and with 3200 indicated horse-power are expected to develop 17 knots. The armament consists of five torpedo-tubes, five 4-inch and six machine guns. In England the *Scout,* the prototype of this class, is a twin-screw torpedo cruiser, 220 feet in length, 34 feet 3 inches in beam, and with 14 feet draught displaces 1450 tons. Like the *Condor* she is subdivided into water-tight compartments and has a steel deck; on her trial she developed with forced draft 17.6 knots and 3350 horse-power. Her armament consists of eleven torpedo-tubes, four 5-inch rifles on central pivots, and eight Nordenfeldt guns.

The *Fearless* is a sister ship to the *Scout*. So highly was the class esteemed that eight others known as the *Archer* class were laid down, and of these the *Cossack, Mohawk, Porpoise, Tartar, Archer,* and *Brisk* have already undergone satisfactory steam trials, while the *Serpent* and *Raccoon* are approaching completion. All these vessels have a protective deck extending throughout their length, and carry a battery of six 6-inch guns on sponsons, two at each extremity, and two in the waist. On the final trials the *Archer* developed under forced draft 17.8 knots and 4122 horse-power, the *Brisk* 18 knots and 3954 horse-power, the *Cossack* 18 knots and 4003 horse-power, the *Porpoise* 17.5 knots and 3943 horse-power, and the *Tartar* 17.28 knots and 3824 horse-power. They have a very low coal consumption, and a coal endurance which was estimated in the *Archer's* case to be sufficient for six days, or 2600 knots at full speed, or for 7000 miles at a 10-knot rate. Both the Russians and the Austrians have vessels of this type, and there is no doubt of the favour with which it is looked upon.

Besides the *Grasshopper* class mentioned in the text, and which includes the *Rattlesnake, Spider, Sandfly,* and *Sharpshooter,* there are two steel cruising torpedo gun-vessels, the *Curlew* and *Landrail,* of 785 tons; these are fitted with a protective steel deck throughout their length, and have a battery of one 6-inch gun, three 5-inch pivots, a supply of machine guns, and four torpedo-tubes. They were intended to develop 14 knots and 1200 horse-power, but on trial the *Curlew* attained 15.081 knots and 1452 horse-power. Owing to a faulty design these ships draw with their proposed weights two feet four inches more water than was expected. In addition to these ships the English have the composite gun-vessels *Buzzard, Swallow, Nymphe,* and *Daphne* of 1040 tons; the *Icarus* and *Acorn,* of the *Reindeer* and *Melita* type; the *Rattler, Wasp, Bramble, Lizard, Pigmy, Pheasant, Partridge, Plover, Pigeon,* and *Peacock,* all of 715 tons displacement, with an average speed of 13.5 knots and from 1000 to 1200 horse-power; and the two despatch and scout vessels *Alacrity* and *Surprise.* The last named displace 1400 tons, and were designed for 3000 horse-power and 17 knots. Both exceeded these expectations, and the *Alacrity* was lately assigned a battery of four 5-inch guns on sponsons, four 6-pounder rapid fire, and two five-barrelled Nordenfeldts. It must be stated, however, that, so far torpedo-boats are not as successful in practice as Admiral Aube would have had the naval world believe.

> Swayed by the concurrent testimony of different officers who conducted or took part in the naval manoeuvres of 1886, professional opinion appears to agree that torpedo-boats are very delicate instruments at best, and that a greater tonnage is imperative where service at sea is anticipated. A day or two in even moderate weather is sufficient to exhaust the staunchest crew on account of the excessive balloting about, and a prolonged voyage has been found to be fatally injurious to the adjustments of the Whitehead for horizontal accuracy. Furthermore, in such small, low craft a correct estimate of the distance, speed, or course of the enemy is most difficult, especially if the officer be in the conning tower, looking through the narrow sight-slits; in anything of a seaway, also, accurate pointing is out of the question. . . . In the course of the past

year Schichau has yielded to Thorny croft the honour of producing the fastest vessel in the world, the owner now being the Spanish Admiralty in place of the Russian. This boat is the *Ariete*, with a speed of 26.18 knots.

It has become a question in the minds of some eminent designers and observers, notably M. Normand, whether or not the extreme speeds sought and obtained in some recent boats are not excessive. Damage to the motive machinery is more to be apprehended than any injury to the hull or casualty among the crew. When it is considered that under ordinary conditions of weather and service the speed of the fastest will be little greater than that of an ordinary twenty-knot boat, the propriety at once suggests itself of devoting to steel plate the extra weight of boiler, water, and engine necessary to produce that practically superfluous horse-power. (Lieutenant Shroeder, U. S. Navy.)

The trials of this year have not confirmed the great promises made for the type by its most able and influential advocates. Many of the English boats broke down, and in few cases were the high speeds realized in actual sea duty. The truth is, torpedo-boats have been brought down to such a condition of refinement to meet the special circumstances of their work that it appears probable they have become too delicate for rough handling. Out of twenty-seven boats that were required to steam a distance of one hundred miles, seven failed to run the course at all, having been, from one cause or another, practically disabled. Such a heavy percentage of failures—one resulted in a loss of life—under a trial test to which the boats might at any time be subjected, arouses a natural doubt as to a policy which is sacrificing for certain impracticable results considerations that are of vital importance.

So far as the French naval manoeuvres proved this year, the torpedo-boats were not equal to the task assigned them. During these experiments a squadron of eight armoured battleships, three cruisers, and two sea torpedo-boats, under command of Vice-admiral Peyron, was supposed to represent a convoy of troop-ships and guard-vessels which was to be intercepted on a voyage from Toulon to Algiers by a torpedo division of four cruisers, one store-ship, and sixteen boats, with the *Gabriel Charmes*, gun-boat, all lying off Ajaccio, under command of Rear-admiral Brown de Coulston.

Vice-admiral Peyron and his heavy squadron left port on the day appointed with a strong northerly gale and a high sea, and shortly after clearing the land the *Indomptable*, an armoured battleship, sustained some damage and had to anchor under the Hyères Islands. The mistral sent the other vessels rapidly on their way to the African coast without slackening speed, all keeping well together, with the two torpedo-boats steaming along under the higher sides of their consorts. On the other hand, the torpedo division of Rear-admiral Brown, which had left Toulon two days before the fictitious convoy, was concentrated at Ajaccio. They ran seaward on Saturday night to find the Peyron ships, but the latter had cleverly given them the go-by in the darkness and bad weather, and the mosquito flotilla was forced to return to Corsica for shelter. Ajaccio was reached by Rear-admiral Brown on Sunday afternoon, and it was not until four-and-twenty hours afterwards that the weather moderated sufficiently to enable him to put to sea again, but by that time the Algiers convoy had already been at anchor in their port of destination since the morning. The preliminary operations were therefore a pronounced failure.

The *Gabriel Charmes* illustrates a design which is similar to that of a torpedo-boat, except that in place of a torpedo tube one 5.5-inch gun is carried forward. The deck is strengthened to bear this weight, and immediately abaft the piece is an armoured conning tower, within which the commanding officer is enabled by an ingenious mechanism to direct the movements of the vessel. The dimensions are as follows: length 132.6 feet, beam 12.6 feet, draught 6.7 feet, and displacement 74 tons. The engines are two-cylindered compound, and developed 560 indicated horse-power and 19 knots. The boats are said to be very cranky even in smooth water, but so highly is their fighting power rated that fifty more have been ordered. In the Mediterranean manoeuvres of May the *Gabriel Charmes* proved to be the swiftest vessel of the torpedo squadron, as on the run from Toulon to Ajaccio she led the others by three hours, and was always in the advance while scouting. One paddle-wheel armoured despatch vessel and seven composite armoured transports complete the record of additions made to the French fleet last year.

The Italian, Russian, German, Austrian and Turkish Navies

The Continental navy next in present interest to that of France is the Italian, owing to the fact that the Italian government, although largely abstaining from the use of armour, has applied itself urgently to developments of gun-power and speed in large warships. The *Duilio* and *Dandolo* were considered in the chapter on the French navy, and their resemblance to the *Inflexible* type pointed out. They are nearly as large as the *Inflexible*, although differing greatly in proportions and form from her. They appear to me to be more objectionable, from the want of armoured stability, if one may so speak, than even the *Ajax* and *Agamemnon*, which are themselves, as we know, more objectionable than the *Inflexible*. The cause of this is to be found in the fact that in designing the British ships, whatever else they may have lost sight of, the Admiralty constructors saw that the more you contracted the length of the armoured citadel, the more necessity there was for giving the ship great breadth. The reason of this can be made clear.

The fractional expression which represents the statical stability of a ship has in its numerator the quantity $y^3 x$, in which y represents the half-breadth of the ship at the water-line, and x the length of the ship. If we regard the stability of the armoured citadel only, and neglect the unarmoured ends, x represents the length of that citadel, and y its half breadth. Now if we take two rectangular citadels, one, say, 100 feet long and 60 feet broad, the other the same length, but only 50 feet broad, then the value of x will be the same for both, but the values of y^2 will be 216,000 and 125,000 respectively, the ship 60 feet broad having, *coeteris paribus*, nearly double the citadel stability of the 50-feet broad ship. On the other hand, if you wish to give the narrower ship the same citadel stability as the broader one, it will be necessary to make her citadel no less than $172^8/_{10}$ feet long. Now the citadel of the *Duilio* is 107 feet in length, and the breadth is 64 feet 9 inches—say 65 feet. I adopt the length from Lord Brassey, who adopts it from Mr. King, but I am inclined to regard it as too small by about five feet, for I observe that in giving the length as 107 feet they give the breadth as 58 feet, whereas they give the breadth of the ship as 64¾ feet. I also observe that they both speak of an "armoured citadel or compartment 107 feet in length," and the word "compartment" seems to point to inside dimensions, and although it seems odd to use these in such a case, it is probable that that has been done. But as there is considerable curvature in the transverse bulkheads, and as the greatest inside length has presumably been given, it may still be practically correct to regard the mean length of the battery as 107 feet. I regret that I have not the means at hand of making certain of the precise length. The citadel of the *Inflexible* is 110 feet long, and its breadth 75 feet, the figures for the *Ajax* being 140 feet and 66 feet, Now presuming the citadels to be rectangular in each case, we shall have,

Inflexible ... y^3x=618,750
Ajax... y^3x =453,024
Duilio... y^3x=452,075

From which it would appear that the *Duilio* of 11,000 tons derives from this element of stability only about as much as *the Ajax* of 8500 tons derives from it, and only about three-fourths of that which the Inflexible of 11,400 had allowed to her. There are other circumstances, of course, which enter into the stability of these ships, but nothing which I know of or can imagine to enable the *Duilio* to compare much more favourably in this respect with the other vessels, deficient as they themselves are. All this applies, of course, solely to the ability of these ships to depend upon their armoured citadels for safety in war: in peace they are all safe enough as regards stability, because they have their unarmoured ends to add largely to it, although I should doubt if the *Duilio* is greatly over-endowed with stability even with her long unarmoured ends intact.

I now come to a series of ships in which the question of the amount of their armoured stability does not arise, because they have no armoured stability at all. For some reason or other Lloyds, in their

THE DUILIO

Universal Register, following bad examples, have arrayed the *Italia* and her successors under the heading of "Sea-going armour-clads." These ships are nothing of the kind, in any reasonable sense of the word, but are, as ships, wholly unarmoured, although carrying elevated armoured towers, and some armour in other places. Mr. King (in his work previously referred to) puts the facts correctly when he says:

> The armour is only used (in the form of a curved deck, be it understood) to keep out shot and shell from the engines and boilers, the magazines, shell-room spaces, and the channels leading therefrom to the upper deck, and to protect the guns in the casemate when not elevated above the battery, and the gunners employed in firing them. But all other parts of the ship above the armoured deck (which is below water, be it said), all the guns not in the casemate, and all persons out of the casemate, and not below the armoured deck, will be exposed to the enemy's projectiles.

Mr. King takes note of this total abandonment of side armour as a means of preserving stability when a ship is pierced at the water-line, and regards this abandonment as a bold defiance of the principles which I have laid down for some years past. I cannot say that I take this view of the matter. I have always discussed this matter from the British navy point of view, and had these ships of the *Italia* type been built for the British navy in substitution of real ironclads, while France, Russia, and other European countries were still building such ironclads, I should have certainly condemned them. The primary requirement of British first-class ships is that they shall be able to close with and fight any enemy of the period whatever, and any defect which unfits them for this work, or makes it extremely dangerous to perform it, is a disgrace to England. Even if armour were given up by other powers, it would be a matter for careful consideration in England whether enough of it for the protection of their existence against contemporary guns should not be retained in her principal ships. England's ability to live as a nation and as the head of an empire is dependent upon her naval superiority, and no price to purchase that can be too great for her to pay. But with Italy the case was and is wholly different. She could not compete with England in naval power, and would not wish to if she could, for she is without an ocean empire to preserve. But Italy has European neighbours, and when she began to build these *Italias* and *Lepantos* she had for neighbour one power, France, which had unwisely persisted for years in building wooden armour-clads, neither strongly protected nor swift, nor very powerfully armed; and I am not at all sure that, to such a navy as France then had, a few extremely fast and very powerfully armed ships such as Italy built were not excellent answers. The *Italia* would have been available also against a very large proportion of the British ironclad fleet, and of the fleets of Austria, Turkey, and Russia. The idea of the Italian ministers clearly was to give weaker ships no time for long engagements with them; but to pounce upon them by means of enormous speed, and to destroy them at a blow by means of their all-powerful ordnance. They might well expect to have with such ships so great a command over the conditions under which they would give battle as to be well able to repair in time, and at least temporarily, such dangerous wounds as they might receive. But more than this cannot be said for such ships: they are not fit to engage in prolonged contests, or to fight such actions as by their assaults on superior numbers and their endurance of close conflict have won that "old and just renown" of which England is so deservedly proud. It seems to me as obvious as anything can possibly be that such ships as the *Italia*, if once adopted as models for other great powers, would admit of easy and cheap answers. Ships of equal speed, merely belted with very thick armour, and armed with an abundance of comparatively light shell-guns, would effectually defy them. There would be no need of enormous and costly armaments, or of ponderous armoured towers, or of huge revolving turrets, for giving battle to ships which any shells would be able to open up to the inroads of the sea, and which, being opened up, would lose their stability, and insist upon turning bottom upward. But for the purposes of the Italian government, as I conjecture them, the *Italia* class of ships, large as they are, have probably been excellent investments, and may continue to be, so long as the priceless value of impregnable belts and interior torpedo defence is understood by so very few.

The Italian government, having completed the *Italia*, is now pressing forward with four other equally large ships (of over 13,000 tons each) of similar type, and with three others of 11,000 tons. Curiously enough, it keeps with these among the "war vessels of the first class" not only the *Palestra* and *Principe Amedeo*, of about 6000 tons, launched in 1871-72, but also the *Roma*, a wooden vessel of 5370 tons, launched twenty years ago, and some four or five iron ships, of 4000 tons and of 12 knots speed, launched more than twenty years ago. I will not occupy time and space by regarding the particulars of these old vessels (having omitted similar ones from my

French tables), but will here give the particulars of the modern vessels of the Italian first class, which alone deserve notice:

MODERN ITALIAN WAR-SHIPS OF THE FIRST CLASS.

Name of Ship.	Displacement.	Indicated Horse-power.	Speed.	Length.	Breadth.	Draught of Water	Greatest Thickness of Armor.	Heaviest Guns carried.
	Tons.		Knots.	Feet.	Feet. In.	Feet. In.	Inches on Sides.	
Duilio......	11,140	7,700	11.1	340	64 4	26 8	22	4 of 101 tons.
Dandolo....	11,200	7,700	11.2	340	64 4	27	22	4 " 101 "
							Inches on Towers.	
Italia.......	13,900	18,000	18	400	74	27 8	19	4 " 103 "
Lepanto....	13,550	18,000	18	400	73 4	27 8	19	4 " 103 "
Re Umberto.	13,250	19,500	18	400	74 9	28 7	19	4 " 106 "
Sicilia	13,250	19,500	18	400	74 9	28 7	19	4 " 106 "
Sardegna ...	13,250	19,500	18	400	74 9	28 7	19	4 " 106 "
Lauria......	11,000	10,000	16	328	67	27	14	4 " 103 "
Morosini....	11,000	10,000	16	328	67	27	14	4 " 103 "
Doria.......	11,000	10,000	16	328	67	27	14	4 " 103 "

The manner in which the towers and guns of the Italia type are arranged is shown in section and in plan, which are taken for convenience from the works of Mr. King and of Lord Brassey, and were prepared, I believe, from official drawings.

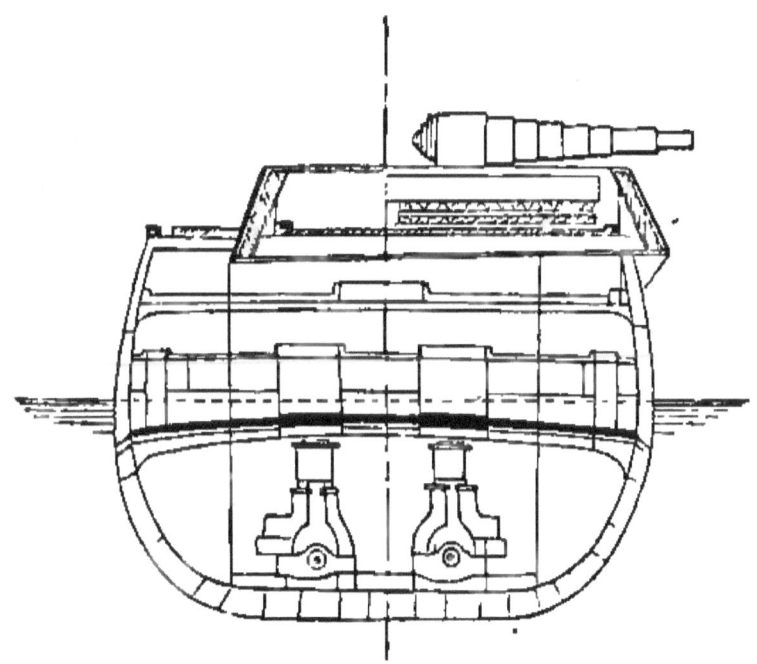

SECTION OF THE ITALIA

Among her unarmoured vessels, in addition to a large number of old and slow small craft, Italy possesses some fast modern warships of

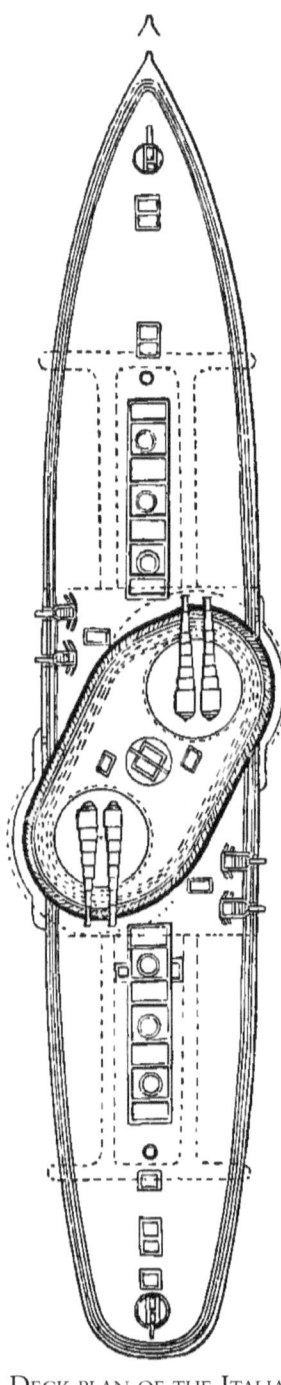

DECK PLAN OF THE ITALIA

the second and lower classes which are deserving of notice. In the first place, she has eight steel vessels ranging from 2500 tons to 3600 tons, which Lloyds describe as "deck-protected cruisers," with a total absence of any justification, I think, excepting that other people have doubtless done so before.

There certainly are people who for business or other purposes would call anything a "protective deck," but why these eight vessels should be removed from the category of unarmoured ships, and constitute a class by themselves, is more than I can imagine even the slightest reason or justification for. I do not know any modern naval gun which will not penetrate an inch steel plate when presented to it as it is in the curving down decks of these vessels. It appears to me a trifling with serious matters to try and induce naval authorities, officers, and seamen to believe that these vessels, and similar ones wherever they are to be found, have any pretensions to be regarded as "protected."

These unarmoured vessels are, however, notable for high speed, three of them being of fifteen knots, and the other five of seventeen knots. One of these 17-knot vessels, the *Giovanni Bausan*, built by Sir William Armstrong & Co., at Newcastle-on-Tyne, so closely resembles the Chilean vessel *Esmeralda* that the engraving of the latter vessel may be taken to illustrate the general character of both. The breadth (42 feet) is the same in each, and so is the draught of water (18½ feet), but the Bausan is a few feet longer than the other. The armament is almost precisely the same, being two guns of

The Italia

about twenty-five tons, mounted one forward and one aft, and six of four tons. I have chosen the *Esmeralda* for the illustration of both vessels because I am in possession of an instantaneous photograph of her at full speed, from which the engraving has been made. This is very interesting, because it exhibits what few readers are likely to have seen, but what most will be glad to see, *viz.*, the form which is taken by the permanent waves that accompany such a ship when steaming at the full speed of seventeen knots in comparatively still water. The engraving also well represents the position of the bow and stern guns.

The 15-knot vessels of Italy are named *Giojà, Amerigo Vespucci, Savoia,* and *Colombo,* of which the *Amerigo Vespucci* is illustrated from a drawing by De Martino. Those of seventeen knots, besides the *Bausan,* are the *Etna, Vesuvio, Stromboli,* and *Fieramosca.* All the last-named vessels carry the same armament as the *Bausan;* the others an armament of 4-ton guns only. The Italian government also possesses (built or building) eight other vessels exceeding or reaching fifteen knots in speed, of which two are built of wood and the remainder of iron or steel. It has likewise of fast torpedo craft a 2000-ton vessel of nineteen knots, which mounts six 6-inch guns and nine 6-pounders; and four others of twenty knots, to carry machine guns, *viz.*, the *Tripoli* and *Goito,* of 741 tons, and the *Folgore* and *Saetta,* of 317 tons. It is also proposed to build six others, of 741 tons and twenty knots, two of which, the *Monzambano* and *Montebello,* have been laid down at Spezzia. They have sixty-two complete first-class torpedo-boats of over one hundred feet in length, and twenty-one second-class, already built, of less than one hundred feet.

It will be seen from the foregoing statement that the Italian navy is one of much importance, capable of working great destruction upon an enemy's fleet of ordinary ships, able to cope with no inconsiderable number of modern vessels, and such as would enable the Italian people and government to speak with a voice that would have to be attentively heeded by any possible ally or any probable enemy in the event of European complications arising, or of a European war becoming imminent. This does great credit to successive Italian political administrations.

Of late the German government has been very active in promoting commercial ship-building and ocean enterprise, but it has been very slack in the development of its imperial navy, and for this reason the Russian navy next claims our notice. Russia, with the continent of Europe interposed between its northern and its southern ports, is compelled to divide its naval strength into two, concentrating one part upon the Baltic and the other upon the Black Sea; and both these di-

The Esmerelda

visions of its navy are under restrictions which approach pretty nearly to the conditions of blockades. With winter comes the natural blockade of Cronstadt and St. Petersburg on the Baltic, and this sometimes lasts so long that I have myself seen the first merchant-vessel of the year approach Cronstadt on the 29th of May, or within a very few weeks of midsummer.

In the South, Sebastopol and Nikolaiev are under the permanent domination of the Bosporus forts and fleets, and of European treaties, which are stronger still. The disasters of the war of 1854 and the political engagements which ensued have also borne heavily upon the naval spirit of Russia, and it says much for the greatness of that country that again, in spite of all these hinderances, it is raising its navy into a position of European importance.

Considering the Black Sea fleet first, the entire interest excited by its armour-clads centres in the three new 16-knot ships, the *Catherine II., Sinope,* and *Tchesme.* These three ships are belted throughout with 18-inch armour, and are armed with six guns of forty tons and seven of four tons, this battery being fought *en barbette* in towers plated with armour fourteen inches thick. The *Universal Register* and the French *Carnet* agree in assigning to the *Catherine II* a length of 320 feet and a tonnage of 10,000, and to the other two ships a length of 314 feet and a tonnage of about 8600. They also agree in describing the horse-power of each of the three ships as 9000 indicated, and the speed as 16 knots. The Admiralty Return previously quoted gives them a speed of 15 knots, and equal tonnages of 10,800 tons. I am unable to give the tonnage decisively, but I know that the tonnage originally intended for these ships was 9,990, and I am in possession of the details of the corresponding weights. The discrepancies as to steam-power and speed are matters of great moment. I believe that both the *Universal Register* and the French *Carnet* are wrong in associating a power of only 9000 horses with a speed of sixteen knots, the fifteen knots given by the Admiralty being the speed expected with 9000 indicated horse-power; but this power is to be obtained with natural draught, while with forced draught the power is to be increased to 11,400, and the speed increased to sixteen knots. The formidable character of these ships needs no comment, although I cannot regard them as nearly equivalent to or as well designed as the somewhat larger *Nile* and *Trafalgar* of the British navy. The only other Black Sea armoured vessels are the slow and small, but somewhat powerful, circular ships *Novgorod* and *Vice-admiral Popoff,* of which the latter is surrounded by 18-inch armour, and carries two guns of

The Amerigo Vespucci

forty tons. A torpedo-vessel of the 600 ton class, developing 3500 horse-power, and 20 knots speed has been built at Nikolaiev.

The Baltic fleet of Russia contains only one finished ironclad of much importance, the *Peter the Great,* of 9340 tons and 14 knots speed, carrying four guns of 40 tons; but two other ships, the *Emperor Alexander II* and the *Nicholas I,* of 8400 tons, *are* now under construction at St. Petersburg. No interest attaches to the *Pojarsky,* the four *Admirals,* and several other old, weak, and slow armour-clads of the Baltic navy. This fleet comprises, however, eight belted cruisers, of which five are important. These are as follows:

Name of Ship.	Displacement.	Indicated Horse-power.	Speed.	Armor.	Principal Armament
	Tons.		Knots.		Guns.
Vladimir Monomach...	5800	7000	15.4	7-inch.	4 of 9 tons.
Dmitry Donskoi.......	5800	7000	16.25	7-inch.	*3 " 29 "
Admiral Nachimoff....	7780	8000	16	10-inch.	8 " 9 "
Alexander Nevsky.....	7572	8000	16	10-inch.	8 " 9 "
Emperor Nicholas.....	8000	8000	16	10-inch.	2 " 40 "

The only fast armoured cruisers of the Baltic fleet are the *Rynda* and *Vitias,* of 2950 tons, 3500 horse-power, and 15 knots speed; and an-other, the *Admiral Kornilof,* now being completed at Nantes, to be much larger and faster. Among-torpedo-vessels there is the twin-screw steel *Iljin,* of 600 tons, which has steamed 20 knots, and carries 19 machine guns; another, of only 140 tons, but to steam 20 knots, has been built at Glasgow; and a third, of like size, but of 17 knots, at St. Petersburg. The torpedo-boats of the Russian navy are given in the parliamentary return as below:

BALTIC TORPEDO-BOATS

Completed: 4 over 100 feet in length; 74 over 70 feet in length; 20 under 70 feet in length. Completed and building: 6 over 100 feet in length, of which 4 are over 150 feet long—total, 104.

BLACK SEA TORPEDO-BOATS

Completed: 5 over 100 feet in length; 8 over 70 feet in length; 6 under 70 feet in length. Completed and building: 7 over 100 feet in length—total, 26.

Russia has also a volunteer fleet consisting of ten vessels of no great fighting value; a Siberian flotilla comprising nine gun-boats and other small craft; a Caspian flotilla of seven small vessels; and an Aral flotilla of still less moment.

In the German armoured navy four citadel vessels figure as having the heaviest (16-inch) armour, but these are of that objectionable *Sachsen* type to which I previously adverted. In order to let the reader see under what slight pretexts some people are prepared to

The Catherine II

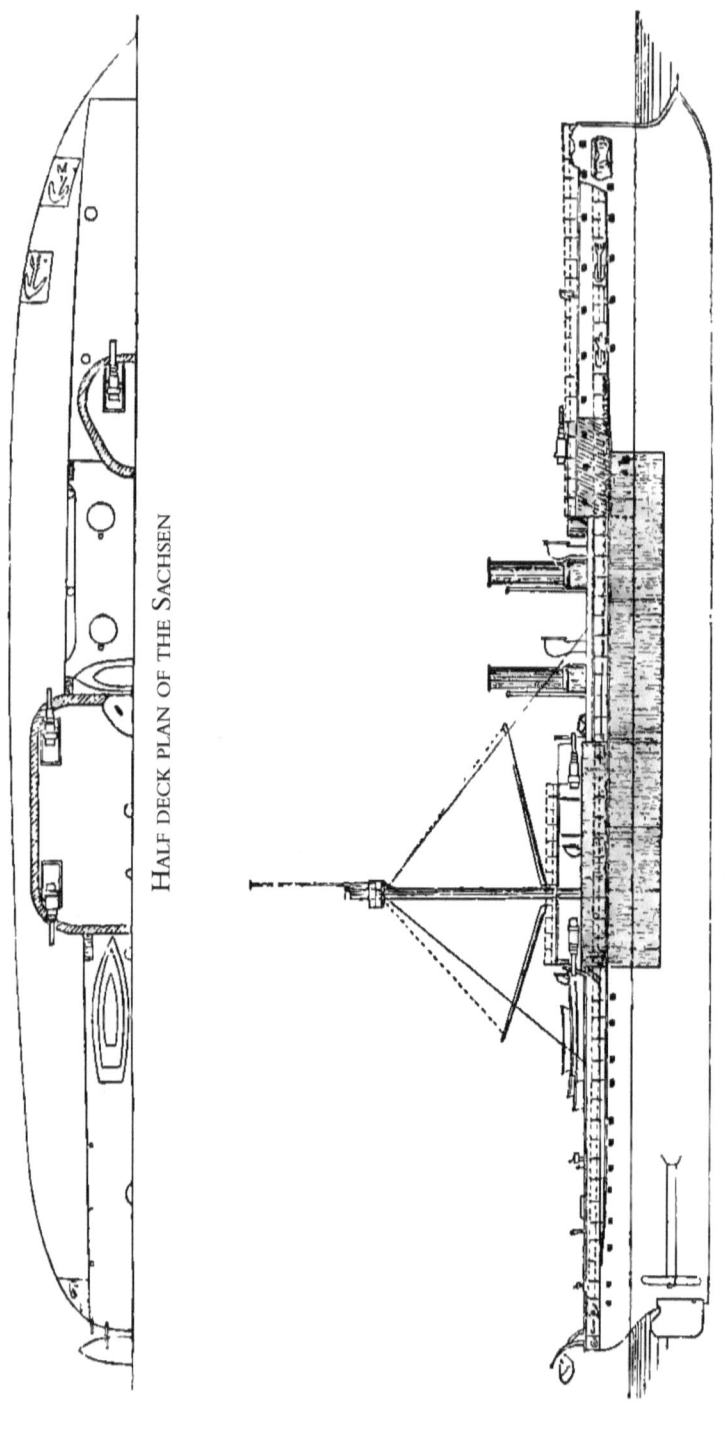

Half deck plan of the Sachsen

Side elevation of the Sachsen

regard ships as powerful ironclads, I give engravings which represent the *Sachsen* in side view and in plan, these illustrations being taken from Captain J. F. von Kronenfels's *Das Schwimmende Flottenmaterial der Seemächte*. The shaded portion in the middle exhibits the extent of this ship's armour; the long white ends are left to depend upon walls of cork, etc., which are very poor—nay, almost imaginary—defences against the effects of explosive shells.

In observing the limitation of the armour in this and similar ships one is tempted to ask, Why stop there? Why not shorten the armour, say to twenty or thirty feet of length, and make it a yard thick, and then enter her in the list of ironclads as a vessel with armour three feet thick? Deck-plating, according to such constructors, is ample for the protection of engines and boilers, and everything else which is below water.

The remaining three ships of this class are the *Baiern*, the *Baden*, and the *Würtemberg*. The engraving of the *Sachsen* represents their general appearance. Their dimensions and other particulars will be given presently in table, but it will be observed that the armament is arranged in a forward and in a midship battery, giving right-ahead fire with four guns, a stern fire with two, and beam fire with three.

The largest ironclad of the German navy is the *König Wilhelm*, of 9750 tons, which steams at 14¾ knots. She is also the most thickly armour-plated (armour, twelve inches); but having been launched eighteen years ago, her guns, although numerous, are only of fourteen tons weight. I designed this ship for his majesty, the late Sultan of Turkey, Abdul-Aziz, but before she was much advanced in construction she was purchased by the Prussian government, and passed from under my care. A few years later I designed the *Kaiser* and *Deutschland* for the Prussian government; and these vessels, built on the Thames, and launched in 1874, although 2000 tons smaller than the *Wilhelm*, steamed but one-fourth of a knot less (14½ knots). They carry 10-inch armour and 10-ton guns. These ships are described further on in this chapter.

The principal ships built in Germany are the *Preussen* and the *Friedrich der Grosse*, which, although designed by the German Admiralty constructors, are but reproductions on a less scale, and with some variations, of the British turret ship *Monarch*. Lord Brassey says:

In the meantime Germany had constructed three turret ships of precisely the same type as the *Monarch*, but of somewhat smaller dimensions. These were the *Preussen*, the *Friedrich der Grosse*, and the *Grosser Kurfürst*. What I do not understand is

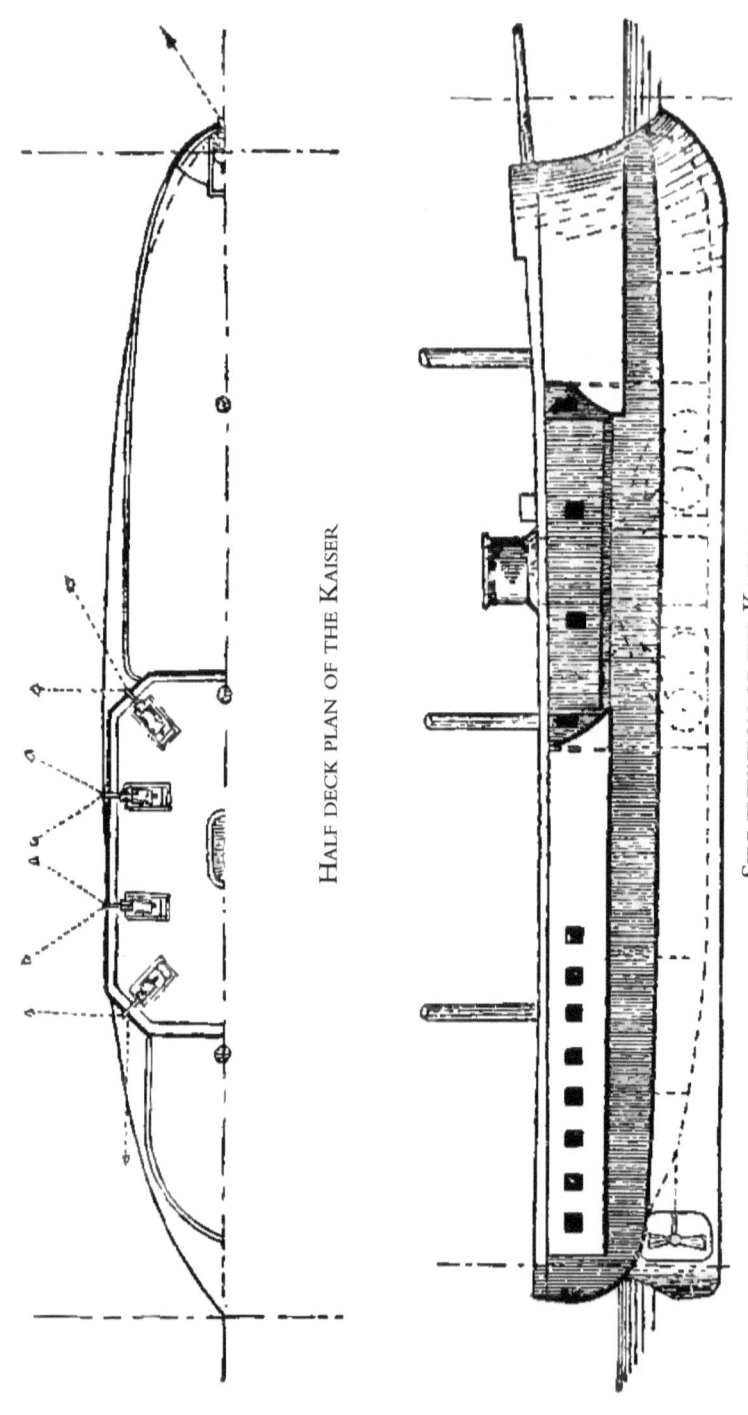

Half deck plan of the Kaiser

Side elevation of the Kaiser

that their armour at the water-line is six inches thicker, while at the turrets it is two inches less, than that of the *Monarch*. (*The British Navy*, vol. i.)

Now, as Lord Brassey elsewhere says:

The *Monarch* is protected with 8-inch armour... and the *Preussen*'s armour-plates at the water-line are 9¼ inches thick, below the water 7¼ inches, and above the water 8¼ inches.

It is obvious that there cannot be the difference of six inches which his first-quoted statement alleges. There doubtless was a difference of an inch, or possibly of two inches, in so far as a few of the armour-plates were concerned, but not more, and how far this difference extended is very doubtful, seeing that nowadays if the constructor of a ship thickens but two or three plates on each side of his ship he feels entitled to speak of her as being armoured with plates of the maximum thickness, and to mislead mankind accordingly. Nor is this surprising, when we see in a late return to the British parliament ships like the British *Collingwood* class, the French *Brennus* class, and the German *Sachsen* class gravely included in the lists of "armoured vessels."

The particulars of the German armoured fleet, leaving out the *Hansa*, a weak and weakly armed ship of only 3500 tons and 12 knots speed, and all smaller armoured craft, are as follows:

SEA-GOING ARMORED SHIPS OF GERMANY.

Name of Ship.	Displacement.	Indicated Horse-power.	Speed.	Maximum Armor.	Principal Armament.
	Tons.		Knots.	Inches.	Guns.
König Wilhelm	9750	8300	14¾	12	18 of 14 tons.
Kaiser	7550	8000	14½	10	8 " 18 "
Deutschland	7550	8000	14½	10	8 " 18 "
Friedrich der Grosse	6600	4930	14	9¼	4 " 18 "
Preussen	6600	4380	14	9¼	4 " 18 "
Baden	7280	5600	14	16	6 " 18 "
Baiern	7280	5600	14	16	6 " 18 "
Sachsen	7280	5600	14	16	6 " 18 "
Würtemberg	7280	5600	14	16	6 " 18 "
Oldenburg	5200	3900	13½	12	8 " 18 "
Friedrich Karl	6000	3500	13¼	5	16 " 9 "
Kronprinz	5480	4800	14¾	5	16 " 9 "

All the above German ships are completed, and have been for a long time, with the exception of the *Oldenburg*, which was not launched until 1884. The *Baden* was launched in 1880, the *Baiern* and *Würtemberg* in 1878, and all the rest earlier—the *Friedrich Karl* and *Kronprinz* nearly twenty years ago, (as at time of first publication). Germany appears to have no ironclad, large or small, under construction at present. It is unnecessary to set forth in detail her small armoured gun-vessels; suffice

THE SACHSEN

it to say that she has one iron turret-ship, the *Arminius*, of 1560 tons, with 7½-inch armour, but only carrying four 9-ton guns, and steaming 10 to 11 knots; and eleven iron vessels of 10 feet draught of water, 1090 tons displacement, 700 horse-power, 9 knots speed, and 8-inch armour, each carrying one 12-inch gun of 37 tons. These were all built at Bremen, and launched between 1876 and 1880, inclusive. They are named after such agreeable creatures as basilisks, crocodiles, salamanders, scorpions, etc., but owing to their small speed would probably prove of less aggressive habits than their names imply. They would nevertheless be very useful in defending the coasts and harbours.

The abstention for the present of the German government, (as at time of first publication), from the construction of armoured ships must not be taken as implying that it prefers the fast unarmoured cruiser as a type of warships, for it has no such cruiser built, and is building but three of very high speed, and one of 16 knots. The particulars of these are as follows:

Name of Ship.	Displacement.	Indicated Horse-power.	Speed.	Armament.
	Tons.		Knots.	Guns.
Elisabeth	4500	8000	18	14 8-inch.
Ariadne	4800	8000	18	14 8-inch.
Charlotte	3360	....	16	....
Loreley	2000	5400	19	2 4-inch.

The Admiralty Return makes no mention of the last ship, as she is but a despatch-vessel, but she is mentioned and particularized in the *Universal Register*. It is to be further observed that the first two vessels on this list are each to have a 3-inch deck, for the protection of the engines, boilers, etc., which fact has induced the Admiralty officers to designate them "protected ships," as they do their own ships of this really unprotected type, and as they have *not* designated the French cruisers *Tage* and *Cecile*. The German navy comprises a few modern and fast frigates, some of which have been honoured with illustrious names, as will be seen from the following list:

GERMAN UNARMORED FRIGATES.

Name of Ship.	Displacement.	Indicated Horse-power.	Speed.	Principal Armament.
	Tons.		Knots.	Guns.
Bismarck	2850	2500	13½	16 of 3¼ tons.
Moltke	2850	2500	13½	16 " 3¼ "
Stosch	2800	2500	13½	16 " 3¼ "
Stein	2800	2500	13½	16 " 3¼ "
Prinz Adalbert	3860	4800	15	{ 2 " 6 " { 10 " 3¼ "
Leipzig	3860	4800	15	10 " 3¼ "
Charlotte	3310	3000	15	18 " 4 "
Gneisenau	2810	3000	15	16 " 3¼ "

There are also some modern corvettes in this navy which may be classed in point of speed with the above frigates; these are:

GERMAN UNARMORED CORVETTES.

Name of Ship.	Displacement.	Indicated Horse-power.	Speed.	Principal Armament.
	Tons.		Knots.	Guns.
Alexandrine	2330	2400	*15	10 of 4 tons.
Arcona	2330	2400	15	10 " 4 "
Carola	2160	2100	14	10 " 4 "
Marie	2160	2100	13½	10 " 4 "
Olga	2160	2100	14	10 " 4 "
Sophie	2160	2100	14	10 " 4 "
Freya	2000	2500	15	8 " 4 "

Regarding the *Alexandrine*—Lloyd's *Universal Register* appears to me to be in error concerning the speed of tills and the next vessel. The *Carnet* gives their speed as fourteen knots, and the Admiralty Return puts it at fifteen knots, which I believe to be the expected speed.

There are about a dozen other smaller and slower gun-vessels and gun-boats in the German navy, but they need not be considered here. As to sea-going torpedo-vessels, the German government took the lead in the production of this type of ship, and had the *Ziethen* launched at Blackwall as a despatch-vessel ten years ago, for a torpedo armament, and with a speed of 16 knots—an example of naval enterprise worth remembering to the credit of Germany. The *Bletz* and *Pfeil*, of 50 *per cent*, larger tonnage, have since been produced in Germany, but only with a speed about equal to the *Ziethen's*. Two torpedo gun-vessels of 855 tons and nearly 2000 horse-power, and 15 knots speed (of which vessels the Admiralty Return makes no mention), were launched at Bremen in 1884. The following is the Admiralty statement as to German torpedo "boats:"

> *Completed*, 58 (43 over 100 feet in length).
> *Completing and building*, 2 torpedo division boats; 30 torpedo-boats over one hundred feet in length.
> Total, 90

Money was voted in 1884-85 for seventy torpedo-boats. When these have been built, the number of German torpedo-boats will be one hundred and five, and these are to be increased to one hundred and fifty.

Reviewing the condition of the German navy as set forth above, it becomes obvious that for some years past the policy of the imperial German government (contrary to that of the Prussian government, which, before the empire, built several large and powerful sea-going ships) has been to avoid all competition in naval matters with the great naval powers, and to apply its moderate expenditure to vessels of a defensive class, such as armoured gun-boats and coast torpedo-boats a policy which, in view of the limited interests of Germany in the Mediterranean and across the seas, has much to commend it.

The Austrian government also, which with less necessity for naval strength now than it had when it possessed Lombardy and Venice, has slackened greatly in the production of ironclads of late years, and has but two, and these of very moderate size, under completion. These are the barbette-battery ships *Kronprinz Rudolph*, of 6900 tons, and the *Kronprinzessin Erzherzogin Stefanie*, of 5150 tons. The former vessel is to carry 12-inch armour, and to be armed principally with three 48-ton guns; and the latter to carry 9-inch armour, and to be armed with two such guns. There is much uncertainty about even the intended speed of these vessels, neither the French *Carnet* nor the *Universal Register* stating the speed, while the Admiralty assigns a speed of 14 knots to the *Rudolph* only. But while the *Carnet* gives the indicated horse-power of each as 6500, the *Register* gives that of the *Rudolph* as 8000, and that of the smaller vessel as much as 11,000. If these latter figures be correct, the *Rudolph* will exceed 14 knots and the *Ferdinand* 16. Austria already possesses two powerful ironclads in the *Custoza* and the *Tegetthoff*, but her *Kaiser, Lissa, Ferdinand Max*, and *Hapsburg* are old wooden vessels, lightly armoured and armed, and need not be further considered. Besides the ironclads already named, she has likewise the three iron central battery and belted ships *Don Juan d'Austria, Kaiser Max*, and *Prinz Eugen*, each of 3500 tons, 2700 indicated horse-power, and 13½ knots speed, with 8-inch armour (the thickest) on the belt, and each carrying eight guns of 9 tons. The unarmoured vessels of Austria (other than those classed as torpedo craft) are numerous, but most of them are small and slow. Those of thirteen knots and upward are but three in number, the *Laudon, Radetzky*, frigates of 3380 tons and 14 knots speed, and the wooden gun-vessel *Hum*, of 890 tons and 13¼ knots speed. Austria is providing herself with several of Sir W. Armstrong & Co.'s light steel vessels of eighteen knots speed for torpedo service, of which she has one, the *Panther*, completed, and two others, the *Leopard* and *Seehund* (all of 1550 tons), under construction. She had also four 14-knot torpedo-vessels, built at Pola and Trieste. Of torpedo "boats" she has the following:

Completed, First class, 135 feet in length, 2; second class, over 100 feet in length, 18; third class, from 85 to 90 feet in length.
Incompleted, First class, 135 feet in length, 2; second class, over 100 feet in length, 8.

Total, 38

The navy of Turkey, which was formidable a few years ago, possessing as it did some of the most powerful and efficient ironclads in the world at that period, both large and small, is rapidly declining in importance in presence of the powerful ships constructed or constructing in England, France, Russia (Black Sea), and Italy. The Turkish navy would not have held its high position so long had it not been for the foresight of the late Sultan Abdul-Aziz, having all his armoured ships built of iron. There is not a wood-built ironclad in the Turkish navy. The largest Turkish armoured ship, and one still very powerful, is the frigate *Mesoodiyeh*, of 9000 tons, built at Blackwall, which in her main features resembles the German *König Wilhelm*, being, like her, of English design, but instead of having eighteen main-deck guns of fourteen tons, she has twelve of eighteen tons, and her battery is consequently of less length. Her speed is fourteen knots. Next to her comes the *Hemidiyeh*, launched in 1885 at Constantinople, of similar type to the other vessel, but of only 6700 tons, and therefore carrying but 9-inch armour, and ten guns of fourteen tons, and steaming at a knot less speed. Turkey has no less than thirteen other ironclads, ranging in tonnage from 2000 to over 6000, in speed from 11 to 14 knots, and in armour from 5½ to 9 inches. The most notable of these, if I may be allowed as its designer to say so, has been the *Feth-i-Bulend (Great Causer of Conquest)*, built at the Thames Iron-works in 1869. This little vessel, although of only 2700 tons displacement, carried a 9-inch armour belt, and a main-deck battery of 6-inch armour protecting four 12-ton guns, placed at the four oblique sides of an octagonal battery, and steamed at four-teen knots—a speed unexampled at that time for an ironclad of her small tonnage. It is a well-known fact that whenever of late years Turkey has had naval work to do, the *Feth-i-Bulend*, on account of her speed, handiness, and general efficiency, was selected by the late lamented Hobart Pasha to perform its most active part.

Of unarmoured vessels Turkey has few worth mentioning as fighting ships, beyond three composite corvettes now under construction at Constantinople, one of 1960 and one of 1160 tons, both of which are to steam at fourteen knots, their armament consisting of eight and six light guns respectively; and one other of 670 tons which is to steam fifteen knots and to carry five light and four machine guns. A steel torpedo-vessel which is to steam at twenty-one knots, and three torpedo cruisers complete the list of new vessels laid down. Turkey has six torpedo "boats "one hundred feet long, built in France; six more of larger size, one hundred and twenty-five feet long, building in Germany; and five of one hundred feet, building in Turkey and France—in all, seventeen torpedo-boats.

This review of Turkish naval force bears out the remark with which I introduced it, and shows that, either from lack of support from the Western European powers or from some other cause, fighting superiority in the Black Sea is being effectually abandoned by Turkey to Russia.

Captain Lord Charles Beresford, R.N., who moved for the Admiralty Return to parliament to which I have made repeated reference, included Greece among the powers whose "fleets" were to be reported on; but as Greece has but two small and weak ironclads, and they are nearly twenty years old, and as she has no other at present even under construction, the pretensions of her "fleet" are scarcely proportional to her political ambitions. She has but one fast cruiser, the *Admiral Miaulis*, and she is only a 15-knot vessel, and carries nothing more in the way of guns than three of six tons and one of five tons. Greece's only "torpedo-vessel" steams no more than fourteen knots, and the Admiralty Return assures Lord Charles Beresford and the world that she has but twenty-seven torpedo-boats, of which seventeen are over and ten under one hundred feet in length, and that she is not building any more. Considering the island interests of Greece and her situation in the Mediterranean, no one can pronounce her naval force as excessive, or regard her government as being tempted to any high heroic policy by her possession of an imposing navy.

I have not mentioned the Spanish or Portuguese "fleets," nor is it necessary to do much more than mention them now. Spain has only one finished ironclad, of over thirteen and less than fourteen knots speed, and that is the *Vitoria*, which was launched at Blackwall, on the Thames, more than twenty years ago. She has thin armour, and could attempt but little in war. Spain is, however, building a large steel turret-ship, the *Pelayo*, of 9650 tons, at La Seyne, to carry two 38-ton and two 48-ton guns, with 18-inch armour on a citadel and 19½ on her turrets. She is to steam at sixteen knots. This one ship will, I presume, when finished, compose the armoured "fleet" of Spain—that country once so great upon the sea. Of unarmoured vessels of war Spain is building several, of which three are to have the advantage of stout steel decks, and one is to be very fast. It will be well to assemble these unarmoured vessels of fourteen knots and upward in a table:

TABLE G.—UNARMORED WAR-VESSELS OF SPAIN.

VESSELS OF FOURTEEN KNOTS AND UPWARD, INCLUDING TORPEDO-VESSELS.

Name of Ship.	Displacement.	Indicated Horse power.	Speed.	Principal Armament.
	Tons.		Knots.	Guns
Reina Regenta	4300	11,000	19	4 of 8 inches.
Alfonso XII	3000	4,400	14	8 " 6 tons.
Aragon (wood)	3300	4,400	14	{4 " 6 " / 4 " 3 "}
Castilla "	3300	4,400	14	8 " 4 "
Navarra "	3300	4,400	14	{4 " 6 " / 4 " 3 "}
Reina Cristina	3000	4,400	14	8 " 6 "
Reina Mercedes	3000	4,400	14	8 " 6 "
Cristabel Colón	1100	1,600	14	3 " 4 "
Don Antonio Ulloa	1100	1,600	14	3 " 4 "
Don Juan d'Austria	1100	1,500	14	5 " 4¾ inches.
Infanta Isabel	1100	1,500	14	5 " 4¾ "
Isabel II	1100	1,600	14	5 " 4¾ "
Velasco	1100	1,600	14.3	3 " 4 tons.
Isla de Cuba	1000	2,200	15	6 " 4¾ inches.
Islas Filipinas	1000	2,200	15	6 " 4¾ "
Destructor (torpedo-catcher)	400	4,000	24	Machine guns.
Alcon (sea-going torpedo-boat)	108	1,200	23	" "
Azor " "	108	1,200	23	" "
Orion " "	88	1,000	20	" "

Spain has likewise four 125-feet torpedo "boats" of 19 knots; one, 105 feet long, of 18 knots; and three or four smaller ones.

Portugal has but one ironclad, central battery type, of 2480 tons, 13½ knots speed, with 9-inch armour, and two 28-ton guns. Of unarmoured vessels she has but three exceeding twelve knots, in speed, *viz*.:

Name of Ship.	Displacement.	Indicated Horse-power.	Speed.
	Tons.		Knots.
Liberal	500	500	16
Zaire	500	500	16
Alfonso de Albuquerque	1100	1360	13.3

All the rest are very slow, and available for little else than harbour defence in time of war.

This concludes our review of the navies of the Continent, The impressions which it has made upon my own mind are mainly these: The minor naval powers are falling more or less completely out of the lists of naval competition. Spain and Portugal have ceased to be, and Greece has not become, of any naval importance—Spain alone making some small effort to keep respectable, but even that effort is chiefly expending itself—as that of the United States government is about to expend itself, by-the-bye—in the production of very fast vessels, which may be useful in preying upon commerce, but which are scarcely fit to fight even pirates, and which a real warship would dispose of with a single round of her battery fire. They will be efficient in running away, no doubt, when danger arises; but "running away" was not the method by which the United States won naval distinction, nor that by which Spain once became great and Greece immortal. The naval policy of Germany is defensive; she is almost without pretensions upon the open sea. Turkey is slowly but surely succumbing to Russia, and in the near future the Russian Black Sea fleet will hold unquestioned mastery over Turkey. Italy has a naval role of her own to play in Europe, and on the whole is playing it well. Austria would do well to hesitate in her present naval condition before again exposing herself to the swift and destructive onslaughts which the tremendously armed and excessively fast Italian ships could make upon her. France is a really great naval power, and there are circumstances which would make a naval conflict between her and England one of the most uncertain in the history of the world. The French have very largely abandoned the protection of their guns by armour; we, most unhappily, have still more largely abandoned the protection of our ships, and it remains to be seen which has been the most foolish. In such a conflict the French would have this advantage over England—the overthrow of their guns, or the destruction of their gunners at them, would not prevent their ships themselves from withdrawing from action and repairing their injuries. What would be-come of our *Ajaxes*, *Agamemnons*, our *Collingwoods* and *Benbows* (both these latter with guns as much exposed as the French, by-the-bye), when their long, fragile ends had been smashed and water-logged, and their high speed consequently gone, is a question which I prefer not to speak of further. There was, there is, there ever will be, but one sound policy for a nation that desires to command the seas, and can afford to do so, and that is to reject all doubtful fads, all dangerous fancies, and to insure without ceasing pronounced superiority in *every known and measurable element of naval power*. New inventions will and must be made; new sources of power, new means of attack, will and must be discovered; but these things take time and money and skill to develop, and that power is the greatest and safest which from time to time and always prefers the thing which must succeed to that which may, and which others fear will, fail in the hour of trial. One hope I, the present writer, have, and it is that the terrible development of the weapons of war—for terrible it is with all its shortcomings—and the enormously increasing cost alike of single actions and of conflicts between squadrons and fleets, will tend to further, and to greatly further, those influences which are happily operating in favour of peace and goodwill among men.

Notes

Italy

The characteristic development of the Italian navy has been the abandonment of side-armour as a protection to stability, and the attempt to obtain high speed and great coal endurance. This bold departure in the matter of armour is due mainly to the fact that Italy's sea policy is governed by conditions which appeal nowhere else with equal force. Sir Charles Dilke writes:

> It is the combination of a large army and a powerful fleet, which really makes Italy formidable; for if Italy has only the fifth army it has the third navy of all the powers. Captain à Court has admirably pointed out how, for a young country, and a country with an overburdened budget, it was not possible to build ship for ship against France, and not within Italy's power to create a fleet numerically equal to that of France, but that it was possible to build a small number of enormous sea-going ironclads of the first class, 'larger, stronger, swifter, and more heavily armed than any afloat.' Were Italy not protected by a powerful fleet, such as might have some chance of holding its own against the French in its own waters, the French fleet could be used to destroy Italian mobilization if Italy had joined an alliance against France. The Italian railway lines could be cut at many places from the coast. Not only from Toulon and Ajaccio, but also from her new port at Biserta, on the Tunisian coast, France could keep watch and could pounce on Italy.
>
> The great difficulty, however, in the way of Italy is caused by her want of coal, for Italy may be said to have no coal for her ships, and the difficulty of getting coal to her southern ports in time of war would be immense if she had not command of the seas. In materially in-creasing the number of her large ironclads Italy has been aiming at nothing less than the command of the Mediterranean as against France; but supposing that France were sufficiently free from the risk of maritime attack elsewhere to be able to concentrate her naval strength in the Mediterranean, it would be a delusion to suppose that the Italian naval forces could hold their own against the French. The Italian material is excellent, no doubt, but the results of Lissa are not encouraging.

To judge from naval expenditure, Italy seems to get a great deal for her money. If we were to look at the figures we should suppose that there were five navies in the world worth counting—the British and French of the first class, and the Russian, German, and Italian of the second class; but as a matter of fact the Russian and German navies are not worth counting by the side of the Italian navy of today. I doubt, however, whether the Italian, German, and Austrian navies could possibly hope to hold the Mediterranean against those of France and Russia, weak as is the Russian navy, in a general Continental war, so high is the estimate which I form of the power of France at sea. Russia, indeed, spends more upon her navy than does Italy; but Russia probably does not get her money's worth. Italy at the present moment, in addition to the two splendid ships which she has at sea, is building or equipping eight first-class sea-going ironclads as against seven being built by France and eleven by ourselves, and she certainly seems to have, as regards the material of her fleet, achieved remarkable results at a low rate of cost.

The Italian fleet, in the event of war, would not have those scattered duties to perform which would fall to the lot of the French and English navies. The fleet of Italy would have to defend the Italian coast against attack, and if possible to keep up the communications with Sicily and Sardinia. Massowah would have to take care of itself, and the Italian fleet would be concentrated, while that of France, in some degree, would have to be dispersed over the whole world; but unless France had to put forth on land such efforts as to need the men and guns of her navy for the defence of her own fortresses, the time of concentration in the Mediterranean would arrive, and a great strain would be imposed upon the Italian fleet.

Those who look upon the Italian navy as being a navy of offence because it consists chiefly of ironclads of the first class capable of holding the seas, forget the necessity imposed upon Italy by her shape and geographical position. It is impossible to defend the coast of Italy by fortifications, and there is no country so vulnerable. The mountains run down the centre of a long, narrow strip, and the strategic railway fines are easily reachable from the sea. On the south, too, Carthage once more threatens Rome. The Italian monster ironclads are certainly not too numerous for the defence of the Italian coast, and in my belief the naval policy which has been pursued by Italy is one which was necessary to her existence, and she is to be congratulated upon the low price at which she has succeeded in obtaining her splendid ships. (*The Present Position of European Politics*)

Owing to this extent and character of the Italian coast, the government believes that absolute safety cannot be secured, and all that may be expected is the disturbance or defeat of any great attempt at invasion or bombardment. This the officials hope to effect by dividing the attention of the enemy's fleet, so that secondary means of defence may be utilized against all attacks. The question, therefore, resolves into one of ships. If armoured vessels had to resist the gun alone, effectual protection, they reason, could be given by increasing the thickness of armour; but since the invention of torpedoes, and the development of great speed in torpedo-boats, the bottoms of ships and not the armoured sides will be the points of successful attack. The best vessels for their needs, therefore, will be such as are capable of making the greatest impression on any given point; that is, such as may be enabled by the partial abandonment of armour to carry enormously heavy guns, and have great speed, the highest coal endurance, and sufficient protection, by new structural devices, to meet without fear any other vessel afloat.

The first fruits of this policy were seen in the central-citadel battle-ships, *Duilio* and *Dandolo*. Apart from their novelty, the mere fact that the Italians could produce such machines with home resources was a surprise to the rest of Europe. The *London Engineer* says:

> The rise of iron ship-building in Italy is almost a romance. It owes its origin to the far-seeing efforts of Italy's greatest statesman, Cavour.... Ten years ago it would have seemed ludicrous to the builders on the Clyde had they been told that a country which had no coal worth speaking of, and whose iron, though abundant, was difficult to get at, and where, moreover, not half a dozen men knew how to do the simplest iron ship-building job, would in the course of those years not only beat them in quality but in price, and would be turning out the largest, the most powerful, and the best built vessels in the world. Such, however, is the case.

Subsequently the Italian Admiralty realized that the ships of the *Duilio* design were deficient in speed and coal endurance, and that their construction forbade the efficient use of a secondary armament for defence against torpedo and other auxiliary boats. So, after much earnest study, the *Italia* type has been adopted. The account in the text needs no amplification here, except to state that in her steam trials she made a maximum speed of 18 and a mean speed of 17.66 knots per hour, although the 18,000 indicated horse-power required by the contract was not developed. Eight of her 6-inch guns, it may be added, have lately been removed.

The *Re Umberto* and *Sicilia* are steel barbette ships, similar to the British Admiral class without the partial armour-belt. Their principal dimensions are, length 400 feet, beam 74 feet 9 inches, mean draught 28 feet 7 inches, and displacement 13,251 tons. The engines of the former are to develop 19,500 horse-power and 17 knots. A complete steel deck three and a half inches thick protects the under-water body. The battery is to consist of four 17-inch 106-ton pair-mounted guns, carried on the fore-and-aft line in two barbettes, which are protected by 18.9 inches of steel armour. There are in addition a number of 6-inch breech-loading rifles, and a supply of rapid-fire and machine guns, and of torpedo-tubes. The *Sardegna*, of the same general type as the *Umberto*, is now being built at Spezzia,

The *Giovanni Bausan*, built at Elswick between 1882 and 1885, is a ram-bowed, schooner-rigged steel cruiser, similar to, but larger than, the Esmeralda, her dimensions being, length 280 feet, breadth 42 feet, draught 18½ feet, and displacement about 3100 tons. She has an underwater protective steel deck one and a half inches thick, and cork-filled cellular compartments about the water-line. The coal supply is 600 tons, the coal endurance 5000 miles at 10 knots, and with 6000 horse-power and 116 revolutions she made on trial a speed of 17.5 knots. Her battery consists of two 10-inch, six 6-inch, and a secondary armament of rapid-fire and machine guns, and of torpedo-tubes.

The steel cruisers mentioned in the chapter, the *Etna, Stromboli,* and *Vesuvio*, are 283 feet 6 inches in length, 43 feet in beam, 19 feet 3 inches in draught, and displace 3530 tons; with forced draft 7700 indicated horse-power and 19 knots are to be developed. Their armament is to consist of two 10-inch (25-ton) Armstrong breech-loaders, mounted in an unarmoured barbette on the fore and aft line, six 6-inch guns on sponsons, eight rapid-fire and machine guns, and four torpedo-tubes—two submerged at the bow and two above water in broadside. The *Fieramosca* of the same class is slightly different in dimensions, and the *Tripoli, Goito, Monzambano,* and *Montebello* are rapid torpedo cruisers, 229 feet 6 inches in length, 25 feet 10 inches in beam, 9 feet 6 inches in mean draught, and of 741 tons displacement. They were designed to develop 4200 indicated horse-power and a speed of 22 knots; but it is claimed that the *Tripoli*, which was launched at Castellamare in August, 1886, realized a speed of twenty-four knots, and maintained a twenty-three knot rate for fifty miles. The engines of the *Monzambano* and *Montebello* will be triple-expansion, and those of the *Tripoli* and *Goito* of the two-cylinder compound inclined type. These vessels have three screws, one shaft coming out underneath the keel at an angle of eight degrees, while the others are carried farther forward on either side. The armament consists of four 57-millimetre and four 37-millimetre rapid-fire guns, of three 37-millimetre revolving cannons, and of five torpedo-tubes, two at the bow, fixed, and three training—one aft and one on each beam. The *Confienza*, a small twin-screw cruiser of nearly the same dimensions, carries four 4.72-inch guns, together with rapid-fire and machine guns, and five torpedo-tubes. She is to develop 17.5 knots and 2800 horse-power, and, like the *Tripoli* class in general, has very light steel frames and plating, and resembles outwardly an enlarged torpedo-boat. The *Folgore* and *Saetta* are torpedo-vessels, similar in type to the *Tripoli*, but smaller; the *Archimede* and the *Galileo* are armed despatch-vessels of the *Barbarigo* type; and the *Volturno* and *Curtatone* are cruising gun-vessels. Other notable additions to the fleet are the partially protected steel cruiser *Angelo Emo*, of 2100 tons, the *Dogali*, and the National Line steamer *America*. The *Angelo Emo* was designed by Mr. White, and built at Elswick for the Greek government, but subsequently she was bought by the Italians, and has, under her new name, made a capital record. The *Dogali* is a twin-screw, lightly protected steel cruiser, built at Elswick. The displacement is 2000 tons, length 250 feet, and beam 37 feet; on the first trial the triple-expansion engines developed 8100 horse-power and a speed of 18.5 knots, and later, with 7600 horse-power and 154 revolutions, a speed of 19.66 knots was attained. The armament is to consist of six 5-inch guns mounted on sponsons—two on the forecastle, two on the poop, and two in the waist.

The *America* is 441 feet 8 inches in length over all, 51 feet 3 inches in beam, 38 feet 5 inches in depth, draws 26 feet aft, displaces 6500 tons, has a coal capacity of 1550 tons, and develops 9000 horse-power and a maximum speed of 17 knots on a consumption of 216 tons of coal per day. She is built of steel, was launched in 1884, purchased in January, 1887, and when refitted is to do duty as a torpedo-depot and transport-vessel. Two iron cruising gun-vessels, the *Miseno* and *Palinuro*, of 548 tons displacement, 430 horse-power, and 10 knots speed, have lately been added to the fleet.

Russia

Russia has shown a marked independence in policy and design. Penned in the Black Sea by treaties, and blockaded in the Baltic for nearly half the year by ice, she has sought in coast-defence vessels, fast commerce-destroyers and torpedo-vessels, the fleet best suited to her necessities. In 1864 a number of monitors, built mainly upon Ericsson's system, were launched, and later four vessels, sea-going, ten-knot turret-ships, were constructed. These are known as the *Admiral* class, and range in displacement from 3754 in the *Lazareff* to 3693 in the *Tchitchachoff*. About 1871 a radical departure was made by the adoption for the Crimean defence of the *circular* or *Popoffka* type. As the shallow waters of this coast forbade the employment of anything normal in design except light, unarmoured gun-boats, recourse was had to a structure of circular form, which with heavy weights could carry a great displacement upon a relatively small draught. Two of these batteries, the *Novgorod* and the *Admiral Popoff*, were laid down, the dimensions of the latter being as follows: extreme diameter 121 feet, diameter of bottom 96 feet, depth of hold at centre 11 feet, extreme draught 14 feet, and displacement 3550 tons. The nominal horse-power was 640, and the number of screws six; the armament consisted of two 41-ton breech-loading guns mounted *en* barbette 13 feet 3 inches above the water-load line, and of four smaller pieces in an unarmoured breastwork. The *Novgorod* attained on her trial eight and a half knots, and the *Popoff* had a mean speed of eight knots.

The Russians were the first to solve the problem of an armoured cruiser in which great speed could be combined with effective protection against the guns of a majority of the high-sea ships then afloat. The *General Admiral*, launched in 1873, and the best known of this class, is built of iron, wood sheathed under water, and coppered. She is 285 feet 9 inches in length, 48 feet 2 inches in beam, and with 21 feet mean draught has 4438 tons displacement. She was designed to steam 13 knots, carry 1000 tons of fuel, and have a coal endurance of 5900 miles at 10 knots; the battery and belt are armoured with six inch plates; the belt is seven feet wide at the water-line, and has, level with its upper edge, a highly curved deck of iron. The type proved so successful that it has been reproduced and improved in most of the great navies.

The *Catherine II, Tchesme*, and the *Sinope* are the most powerful battleships of the Russian fleet. The first and second were launched in May, 1886, the third in June, 1887; they are built of iron and steel (wood sheathed and coppered), have ram bows, and are of the following dimensions: length 339 feet, beam 69 feet, mean draught 26.5 feet, displacement 10,181 tons. They are encircled by a belt of compound armour twelve to eighteen inches thick, and have a complete 3-inch protective deck. Within a 14-inch armoured pear-shaped redoubt six 12-inch rifles are pair-mounted on Moncrieff disappearing barbette carriages; seven 6-inch guns are carried on the gun-deck—six in broadside and one on a shifting pivot mount—and the secondary battery is composed of seven Hotchkiss revolving cannon and seven torpedo-tubes. The engines of the *Catherine II.* and the *Tchesme* are of the vertical compound three-cylinder type, and are to develop 11,000 horse-power and 16 knots; the engines of the *Sinope* are of the triple-expansion type, and are to develop 10,000 horse-power with natural, and 13,000 with forced, draught. The cost of each vessel will be about $4,500,000, The second ship of the *Emperor Alexander II.* type, now building at St. Petersburg by the Franco-Russian Company, and named the *Nioholas the First.*, is to be 8440 tons in displacement, 327 feet in length, 67 feet in beam, and have 25.5 feet draught. These ships carry a complete belt of steel armour six to fourteen inches thick and nine feet wide, and a curved steel deck, three inches thick, covers their under-water bodies. The battery is to consist of two 12-inch guns, mounted in a pear-shaped barbette tower forward; in the broadside there are to be four 9-inch, eight 6-inch, and four 3.5-inch rifles, together with a number of Hotchkiss guns. The barbette tower has steel armour, ten inches thick, and the usual torpedo-tubes are to be supplied. The estimated horse-power is 8000 and the speed 16 knots.

The *Vitias* and *Rinda*, steel cruisers, in which the vital parts will be covered by a curved steel deck one and a half inches thick, are of 2965 tons displacement, develop 3000 indicated horse-power, and have a speed of 15 knots. The *Pamjatj Azowa*, a cruiser of the *Imperieuse* type, with a partially armoured belt and barbette batteries, is expected to develop 8000 indicated horse-power and 17 knots.

The rapid, unarmoured steel cruiser building at St. Nazaire, and named the *Admiral Korniloff*, is of 5000 tons displacement, has triple expansion engines, a curved steel deck to protect the machinery and boilers, and a cellular subdivision, which it is hoped will insure stability in case of perforation at or below the water-line. For the Black Sea fleet six heavy gun-vessels have been projected; these are the *Uralets, Tereto, Kubanets., Zaporojets, Donets,* and *Chernomorets*, of 1224 tons displacement and 2000 horse-power; their armament is to be two 8-inch guns, one 6-inch breech-loading gun, two 6-pound rapid-fire pieces, four revolving cannons, and two torpedo-tubes. The *Bobr* and *Sivoutch* are heavily armoured gun-vessels of a new type; the *Coreets* and *Manchooria* are small twin-screw cruisers of 1213 tons displacement, and the *Aleuta* is a transport, the interior arrangements of which are designed mainly for the storage and distribution of high explosives and torpedoes.

The remarkable development of machine-gun fire on board torpedo-vessels is shown in the *Iljin* and the *Saken*, a type which occupies the middle ground between the smaller class of French torpedo-cruisers and the British torpedo-boat catchers. The *Iljin* carries twelve revolving cannons and seven Hotchkiss rapid-fire guns, and has seven above-water torpedo-tubes, one on each side of the stem, one on the port side of the stern, and four in broadside. Russia has a most effective fleet of torpedo-boats, some of which have attained very high speeds under the usual test conditions of carrying 14½ tons of ballast, coal sufficient for 1200 miles, and a crew of eighteen. The Russian officers have already shown their skill and daring in this system of warfare, and, should they be called upon, there is no doubt that the whole capacity of these boats will be tested under the guidance of a courage and an intelligence which are unsurpassed in any other navy of the world.

Spain

On January 12, 1887, a new naval programme was announced by the Spanish government, and the following types and numbers of vessels were designated as necessary for the modern fleet:

1. Eleven protected steel cruisers: eight to be of 3200 tons, and three of 4500 tons displacement. The armament will be of the 9.45 or the 11-inch calibre Hontoria breech-loading guns, mounted on central pivots, with smaller pieces in broadside and a secondary battery of rapid-fire guns and torpedoes. All the ships are to be constructed on the cellular system, with double bottoms and water-tight compartments, are to have triple expansion engines and twin screws, and are expected to attain a speed of 19 knots with natural, and 21 knots with forced draft.

2. Six steel torpedo-cruisers of 1500 tons displacement and a speed of 23 knots. They are to mount central pivot guns from 6.3 to 7 inches in calibre, in addition to a number of smaller broadside guns, revolving cannons, and torpedoes.

3. Four torpedo-cruisers of 1100 tons displacement, to develop a speed of from 18 to 21 knots, and to be furnished with a heavy primary and the usual secondary battery.

4. Twelve steel torpedo gun-boats, six to be of 600 tons displacement, and six of 350 tons, with a speed of not less than 16 knots.

5. Sixteen steel torpedo gun-boats of 200 or 250 tons displacement, with a speed of 14 to 16 knots.

6. Ninety-six torpedo-boats, 100 to 120 tons displacement, with a maximum speed of 24 knots, and a coal endurance of 1500 miles.

7. Forty-two torpedo-boats of 60 to 70 tons displacement.

8. One transport of 3000 tons, to be equipped as a floating arsenal or machine-shop.

9. Twenty steel steam-launches of from 30 to 35 tons displacement, built on the lifeboat system, and fitted with triple-expansion engines, to drive the boats from 12 to 14 knots per hour.

The cost of the new fleet will be:

For new vessels...............................	189,900,000	Pesetas.
For completing vessels now building..........	22,600,000	"
For arsenals.................................	10,000,000	"
For submarine defence.......................	2,500.000	"
Total.............	225,000,000 *	"

When these vessels are finished Spain will have the following fleet:

New Vessels.		Present Vessels.	
Armor-clad...............................	1	Armor-clads...............................	2
First-class cruisers.......................	12	First-class cruisers.......................	6
Second-class cruisers.....................	13	Second-class cruisers.....................	16
Torpedo gun-boats	32	Smaller vessels...........................	37
First-class torpedo-boats..................	100	Total afloat	61
Second-class torpedo-boats................	50	To be constructed	229
Steam-launches...........................	20	Grand total of fleet	290
Arsenal transport.........................	1		
Total.............	229		

In the Reina Regents the Spanish government expects to have the fastest cruiser afloat. Her keel was laid in the Thompson's Yard at Clydebank on June 11, 1886, and she was launched February 24, 1887. She is built of steel, is 320 feet in length, 50 feet 7 inches in beam, has a sea-going displacement of 4800 tons, and with her full capacity of coal and stores on board, a displacement of 5600 tons. The motive power consists of two independent, horizontal, triple-expansion engines (each working in its own compartment), which are capable of developing with forced draft 12,000 indicated horse-power and a speed of 20.5 knots. The battery will consist of four 9.45-inch Hontoria rifles, mounted on platforms raised four feet above the deck, and situated two forward and two abaft the superstructure; of six 4.72-inch Hontoria guns, two mounted each side on sponsons, and one each side in a recessed port; of eight 6-pounder rapid-fire guns, six revolving cannons, and five above-water torpedo-tubes. The ship has a complete steel deck, curving from about six feet below the water-line to its horizontal height; this latter section is about one-third the width of the ship, and is three inches thick over the engines and boilers, and one inch thick for the rest, while the inclined and curved sides are four and three-quarter inches thick. To assist in excluding water when pierced, a complete belt of cellulose extends around the ship inside the inner skin, and about the height of the water-line.

The torpedo-boat chaser *Destructor* is not only a good sea-boat, and capable of making a long passage at high speed, but has proved herself to be one of the fastest vessels afloat. She has 350 tons normal displacement, and when fully loaded and equipped 458 tons. Her engines have developed 3829 indicated horse-power. During ten days in November, 1886, a maximum speed of 23¾ knots was attained, and on December 13th of that year she reached a mean speed for four hours of 22.65 knots, and an estimated coal endurance of 5100 miles at 11½ knots, and of 700 miles at full speed. In January, 1887, she ran in twenty four hours from Falmouth to Finisterre, thus covering the 495 miles at a mean speed of 21 knots.

The *Pelayo*, a barbette ship of the *Amiral Duperré*, class, has a complete water-line belt of steel, 6 feet 11 inches wide, and from 11.8 to 17.72 inches thick. The steel armour on the barbette towers is 11.8 inches, and the protective deck which extends throughout her length is 3.5 inches thick. The dimensions are as follows: Length 344 feet 6 inches, beam 66 feet 3 inches, draught 24 feet 8 inches, and displacement 9902 tons. The armament consists of two 12.6-inch 48-ton guns in the barbettes; of two 11-inch guns on sponsons, one each side; of twelve 4.72-inch guns in broadside, and of one 6.3-inch piece in the bow. The secondary battery is composed of fourteen rapid-fire and machine guns and seven torpedo-tubes. The contract horse-power is 7000, the speed 15 knots, and the coal endurance (the supply being 700 tons) is sufficient for 885 miles at 15 knots, and 2340 miles at 13 knots. The five ships of the *Infanta Isabel* class are launched, and the small steel cruisers *Isla de Luzon* and *Isla de Cuba* are rapidly approaching completion.

Austria

Austria has under construction this year the two armour coast-defence vessels described in the text: the *Tiger*, a 3800 ton protected cruiser of the latest type, and the *Meteor*, a torpedo cruiser of the *Leopard* and *Panther* class. These last-mentioned important additions to the fleet are 224 feet long, 34 feet beam, 14 feet draught, and of 1550 tons displacement. They differ from the English *Condor* and the French *Archer* in these particulars: first, the steel protective deck is not continuous; secondly, the engines are of the vertical, inverted, triple-expansion type; and thirdly, the engine cylinders are protected by steel shields surrounded by coal or sand-bags. The armament consists of four large-range Krupp guns, mounted in sponsoned turrets, of numerous machine and rapid-fire pieces, and of four above-water torpedo-tubes. Under natural draft 17.6 knots, and with forced 18.9 knots, were accomplished.

The 87-ton torpedo-boats *Falke* and *Adler*, built by Messrs. Yarrow & Co., are 135 feet long, with 14 feet beam, 5½ feet draught aft and 2¼ feet forward. The engines are of the three-cylinder, compound, surface-condensing type, and developed 1250 horse-power and 22.4 knots in fighting trim. The coal supply of twenty-eight tons is expected to give an endurance of two thousand miles at ten knots. Their armament is composed of two machine guns and two torpedo-tubes, which discharge straight ahead. The *Habicht*, a 90-ton torpedo-boat, built by Schichau, was designed to develop with a load of 14½ tons a speed of 20½ knots, and to have a coal endurance of 3500 miles at a 10-knot rate; but on trial she realized 21 77 knots for three hours. It is understood that future boats will be much larger, approaching 300 to 400 tons displacement. The budget for 1887 provides 720,000 florins for torpedo-boats and vessels.

Though Austria holds a secondary place as a maritime power, she is, of all the Continental nations, the one most liable to precipitate the next great war, and it seems strange, therefore, that she does not try to ac-quire a great number of those special classes of ships which, after all, are the only logical answers the weaker naval countries can make to the more powerful.

> While the Austrian military position, in spite of the desire of the emperor for military reform, is still weak, I cannot find words too strong to praise the political ability with which the Austrian empire is being kept at peace and kept together. The Austrian empire is a marvel of equilibrium. The old simile

of a house of cards is exactly applicable to its situation; and just as in the exercises of acrobats, when seven or nine men are borne by one upon his shoulders, it is rather skill than strength which sustains them; so, if we look to the Austrian constitution, which we shall have to consider in the next paper in this series, it is a miracle how the fabric stands at all. At the same time it is impossible for Austria, although she can maintain her stability in times of peace, to impose upon either her Russian or her German neighbours as to her strength for war. Prince Bismarck is obliged, with whatever words of public and private praise for the speeches of the Austrian and Hungarian statesmen, to add the French and Russian forces together upon his fingers, and to deduct from them the Austrian and the German, with doubts as to the attitude of Italy, doubts as to the attitude of England, and contemptuous certainty as to the attitude of Turkey.

If Austria could have presented Prince Bismarck not only with an English alliance, but with an English, Turkish, and Italian alliance, he might possibly have allowed her to provoke a general war; but with the difficulties attendant upon a concession of territory to Italy, except in the last resort, and with Turkey at the feet of Russia, it was difficult for Prince Bismarck to go further than to say to Austria, 'Fight by all means, if you feel yourself strong enough to beat Russia single-handed. France and Germany will "see all fair," and you can hardly expect anybody effectually to help you.' Prince Bismarck deals with foreign affairs on the principles upon which they were dealt with by King Henry VIII. of England, when that king was pitted against the acutest intellects of the empire and of France. His policy is a plain and simple policy, and not a policy of astuteness and cunning, and almost necessarily at the present time consists in counting heads. (Dilke)

There have been no additions of any importance to the fleets of the other European powers since the publication of Sir Edward Reed's article, and their policy has in no way been changed from that epitomized in the text. The apathy of Germany is inexplicable, and as for the others, there seem, except with Turkey, perhaps, no good reasons why they should strive to create fleets, as they are either too poor to build and support them, or their dangers from maritime attack are not great enough to make a large navy necessary.

Holland has lately launched the *Johom Willem Friso*, which is the last of six large cruisers:

> ... of which the others are the *Atjeh, Tromp, Konigin Emma der Nederlanden, De Buyter, and Van Speyk*. ... All these vessels are built of iron and steel, sheathed with wood to four feet above the water-line, and coppered. They are of 3400 tons displacement and of the following dimensions: Length 262 feet 5 inches, beam 39 feet 4 inches, and mean draught 18 feet 4 inches. Their armament is six 6.7-inch Krupp guns (one carried in the bow, one in the stern, and the others in broadside), four 4.72-inch Krupp pieces in broadside, six 37-millimetre revolving cannons, and a supply of Whitehead torpedoes. The engines drive single screws, and have an estimated horse-power of 3000, which has been slightly exceeded by some and not attained by others. The speeds vary from 14.1 knots to 14.7 knots. The coal supply is 400 tons—sufficient for six and three-quarter days' steaming at full speed or for thirteen days at ten knots. (Lieut. Colwell, U.S.N., in *Recent Naval Progress*)

Denmark has the *Valkyrien*, a steel cruiser of 2900 tons, fitted with a good battery and five torpedo-tubes, and designed to develop 5000 horse-power and 17 knots. Her new double-turreted, armoured, coast-defence vessel *Iver Hvitfeldt* has developed a maximum speed of 15.6 knots. From data furnished by First-lieutenant Tasker H. Bliss, U. S. Artillery, the peace strength of the principal Continental nations may be summarized as follows:

Country	Permanent Establishment				Cost.		
	Population.	Army.	Navy.	Total.	Army.	Navy.	Total.
Germany..	45,234,000	427,274	11,109	438,383	$86,000,000	$10,000,000	$96,000,000
Russia ...	78,000,000	760,000	29,008	789,000	146,000,000	20,500,000	166,500,000
Austria...	38,000,000	287,000	8,500	295,500	58,500,000	3,500,000	62,000,000
Italy.....	29,000,000	210,373	15,055	225,428	41,000,000	10,000,000	51,000,000
France ...	37,672,000	518,642	43,235	561,877	119,250,000	41,000,000	160,250,000
England ..	35,000,000	199,000	58,000	257,000	78,000,000	53,750,000	131,750,000

A rough analysis of these figures shows that in strength of army Russia is first, France second, Germany third, Austria fourth, Italy fifth, and England sixth; and that in naval strength England is first, France second, Russia third, Italy fourth, Germany fifth, and Austria sixth. The cost of each nation's navy is in direct proportion to its strength of *personnel*; but in armies England, though last in numbers, changes place with Italy, which supports its forces with the least expenditure. It may be added that in total cost England, with next to the smallest force, pays more than Germany, with the third largest in numbers.

The percentage of expenditures is as follows:

Service per Capita of Population.				Cost per Man.		Taxation per Inhabitant to Support the Peace Establishment.
Country.	Army.	Navy.	Total.	Army.	Navy.	
Germany...	0.94	0.02	0.96	$201.00	$900.00	$2.12
Russia.....	0.97	0.04	1.01	$192.00	$707.00	$2.13
Austria....	0.77	0.02	0.79	$204.75	$411.75	$1.65
Italy......	0.73	0.05	0.78	$194.75	$664.25	$1.75
France.....	1.37	0.11	1.48	$230.00	$931.00	$4.25
England...	0.57	0.16	0.73	$391.50	$924.75	$3.76

An examination of *Lloyd's Universal Register of Shipping* for 1887 shows that the present condition of European navies may be popularly-stated in this manner: England has 6 guns capable of penetrating 36 inches of unbacked iron, and 16 others which can penetrate 28 inches of the same material; Italy has 20 guns which can penetrate 33 inches of iron; France has 14 guns which can pierce 27 inches, and 14 others able to penetrate 25 inches of unbacked iron. Russia has 20 guns and Spain 2 which can pierce 24 inches of iron. No other power has any guns capable of equivalent results. In other words, of guns able to penetrate 24 inches of unbacked iron, France has 28, Italy 20, Russia 20, Spain 2, and Great Britain 22.

In warships of 20 knots and above, England has 1, France 1, Italy 10, Spain 2, and other European nations 4; of 19 knots speed, England has 11 ships, France 10, Germany 3, Italy 2, and other nations 9; of 18-knot ships, England has 5, France 7, Germany 2, Italy 6, and other nations 6. English supremacy is, however, chiefly seen in 17-knot ships, of which she has 25, mounting 181 guns; France, 4 with 20 guns; Italy, 5 with 40 guns; and other nations 4 with 19 guns. England has 11 ships of 90 guns that can steam 16 knots, whereas France has 3 only of 58 guns. At 15 knots, France has 16 ships of 214 guns, and England 12 ships of 126 guns; and at 14 knots, France has 28 ships of 334 guns, and England 15 ships of 252 guns. Summarizing these figures, it appears that with speeds above 14 knots England has 80 ships of 795 guns, France 69 of 699 guns, Germany 35 of 285 guns, and Italy 41 of 201 guns.

Out of a total mercantile tonnage now afloat of 20,943,650, Great Britain and her colonies own 10,539,136. The total steam mercantile tonnage of the world is 10,531,843, and of this Great Britain and her colonies own 6,595,871, or nearly two-thirds of the whole.

The United States Navy in Transition
by Rear-Admiral Edward Simpson, U.S.N.

The condition of the navy of the United States is not such as any citizen of the country would desire. Pride in their navy was one of the earliest sentiments that inspired the hearts of the people when the United States took their place as a nation, and the memory of its deeds has not faded during the subsequent years of the country's aggrandizement. Time was when that section of the country most remote from the sea-coast was indifferent to it, owing to the more immediate demand on its attention for the development of internal resources; but the rapid settlement of our Western lands, and the annihilation of distance produced by rapid communication, have tended to preserve the unity of interests of the separate sections, and the happy system that obtains through which officers are appointed to the navy keeps it an object of personal concern to all the States of the Union.

The present condition of the navy is not such as to satisfy the desire of the people that it should be sustained on a footing commensurate with the position of the nation, and in keeping with its ancient reputation. For many years circumstances have intervened to prevent a judicious rehabilitation of the navy, notwithstanding that its needs have been faithfully presented to Congress year after year. The country has been wonderfully favoured with peace at home and abroad, and no urgent call to arms has roused the nation to prepare for war. The rapidity with which a large fleet of cruising ships for blockading purposes was extemporized during our civil war has left a hurtful impression on the public mind that in an emergency a simi-

U.S. SIDE-WHEEL STEAMER POWHATAN

lar effort might prove equally efficacious—disregarding altogether the difference in circumstances of contending with an enemy possessed of a naval force and with one possessed of none. The economists have suggested that as all that relates to ships, guns, and motive forces was being rapidly developed by others, it would be a saving of the people's money to await results, and to benefit by the experience of others; and, again, party rivalry and contentions have assisted to postpone action.

It has never been the intention that the navy should die from neglect and be *obliterated*. Yearly appropriations have been faithfully passed for the support of the personnel, and for such repairs as were found to be indispensable for the old ships that have been kept in commission; but it is now seen that this system of temporizing has been the poorest kind of economy. This money has been invested necessarily in perishable material, the amounts have been insufficient to compass new constructions, whether in ships or guns, and the only use that could be made of them was to repair wooden ships and convert cast-iron guns, whereas the work needed was to construct steel ships and to fabricate steel guns.

In referring to the navy of the past, it is impossible to avoid recalling the feeling of pride with which an American seaman—officer or man—walked the deck of his ship. This feeling was common to the naval and commercial marine. Our wooden ships that sailed the ocean from 1840 to 1860 were the finest in the world. The old frigate Congress in 1842 was the noblest specimen of the frigates of the day, and the sloop of war *Portsmouth* was unsurpassed as a corvette. The clipper ships of that period need no eulogy beyond their own record. These ships were the models for the imitation of all maritime nations, and among the constructors of the period can be recalled, without detriment to many others omitted, the names of Lenthall, Steers, Pook, and Delano. The poetry of sailing reached its zenith during this period.

But there is no sentiment in progress; its demands are practical and imperative, and the great motive power, steam, was being crowded to the front even during this the greatest development in the era of sails. Advanced ideas could not be resisted, and steam was admitted as an auxiliary; but our development in naval construction still stood us in good stead, and enabled us to supply ships with auxiliary steam-power, which continued to be prominent for many years as standards to which others found it to their advantage to conform.

Before the final abandonment in the navy of sailing-ships, pure and simple, an effort at a compromise was made by limiting steam to side-wheel vessels, and a number of fine ships were built in the forties

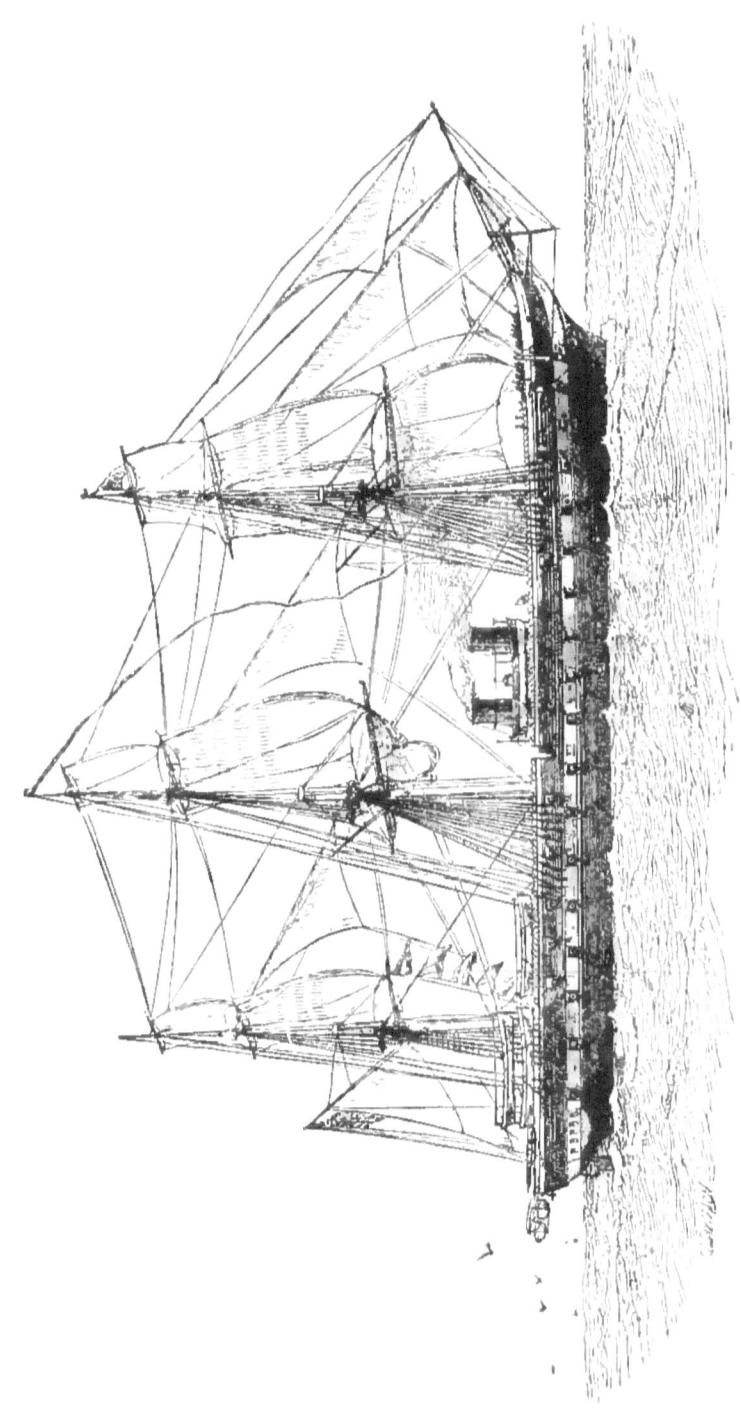

U. S. FRIGATE FRANKLIN, OF THE MERRIMAC CLASS

which did good service, and were a credit to the country, answering as they did the demands of the time. The *Mississippi, Missouri, Susquehanna, Saranac*, and *Powhatan* carried the flag to all parts of the world for many years, some of them enduring to bear their share in the late war, while the *Powhatan* was borne on the list of vessels of the navy until within a few months.

This vessel was built at Norfolk, Virginia, in the year 1850. Her length was 250 feet, beam 45 feet, draught of water 19.6 feet. She had a displacement of 3980 tons, and attained a speed of 10.6 knots per hour with an indicated horse-power of 1172. The capacity of her coal-bunkers was 630 tons. Her battery consisted of sixteen 9-inch smooth-bore guns. She was built of seasoned live-oak, and though frequently under repairs, retained so much of the strength of her original construction that she escaped the sentence of condemnation until recently, (as at time of first publication).

The *Princeton*, of great fame, and the *San Jacinto*, were the only ships with screw-propellers that appeared in the period under consideration, the screw then being considered of such doubtful propriety as to need the test of tentative experiments. These ships have long since disappeared, but the screw remains, and side-wheels are mainly relegated to boats for inland waters. Confidence being established in the screw-propeller, construction on the principle of auxiliary steam-power was decided on, and ships of different classes were added to the navy in such numbers as the varied duties required.

There were those at that time who, wise beyond their generation, recognized the full meaning of the advent of steam, and saw that it must supplant sails altogether as the motive power for ships. These advocated that new constructions should be given full steam-power, with sails as an auxiliary. But the old pride in the sailing-ship, with her taunt and graceful spars, could not be made to yield at once to the innovation; nor could the old traditions pointing to the necessity of full sail-power be dispelled; so it was considered a sufficient concession to admit steam on any terms, and thus the conservative and temporizing course was adopted of retaining full sail-power, and utilizing steam as an auxiliary.

The United States government was not alone in this policy. It was the course pursued by all other maritime nations, and for some years the United States retained the lead in producing the most perfect types in this new phase of naval construction.

In 1851: Congress passed an act authorizing the construction of the *Merrimac* class of frigates. The famous ships immediately built un-

U. S. SLOOP-OF-WAR HARTFORD

der this act were the *Merrimac, Wabash, Minnesota, Roanoke*, and Colorado. All of these vessels got to sea during 1856 and 1857, and were followed, at an interval of ten years, by the *Franklin*, which was a larger ship, and an improvement on the original type.

The *Franklin* was built at Kittery, Maine. Her length is 266 feet, beam 54 feet, draught of water 24 feet. She has a displacement of 5170 tons, and attains a speed of 10 knots per hour with an indicated horse-power of 2065. The capacity of her coal-bunkers is 860 tons. Her frames are of seasoned live-oak, and she is in use as a receiving-ship.

The *Merrimac* was the first vessel of this type which got to sea. She was sent to European waters, and on her arrival in England, early in 1856, she became at once the object of the closest scrutiny, resulting in the unqualified approval of foreign naval architects. The English Admiralty proceeded to imitate the type, and many keels were soon laid in order to reproduce it. The ships built after this model were the crack ships of the time in the English navy, and carried the flags of the commanders-in-chief of fleets.

In 1858, 1859, and 1860, the *Hartford* class of large corvettes appeared. These are full-rigged ships. The class comprises the *Hartford, Brooklyn, Pensacola, Richmond*, and *Lancaster*.

The *Hartford* was built at Boston in 1858. Her length is 225 feet, beam 44 feet, draught of water 18.3. She has a displacement of 2900 tons, and attains a speed of 10 knots per hour with an indicated horse-power of 940. The capacity of her coal-bunkers is 241 tons. Her battery consists of one 8-inch muzzle-loading rifle (converted) and 12 9-inch smooth-bores. These ships were built of live-oak, and endure to the present day, (as at time of first publication). They were reproduced by England and France when they made their appearance, and are now, except the *Trenton*, the only ships in service which can accommodate a commander-in-chief of a squadron. They are kept constantly employed showing the flag abroad, but it is with difficulty that they are retained in suitable repair for service.

This class of ships has good speed under sail, with the wind free, but their light draught prevents them from being weatherly on a wind. Much of their cruising is done under sail, which tends to lengthen their existence. Under the late act of Congress prohibiting repairs on wooden ships when the expense shall exceed twenty *per cent*, of the cost of a new vessel, these ships must soon disappear from the navy list. When that time shall arrive, and steel cruisers shall be substituted, the name of the *Hartford* should be preserved as closely associated with the glory that Farragut shed upon the navy.

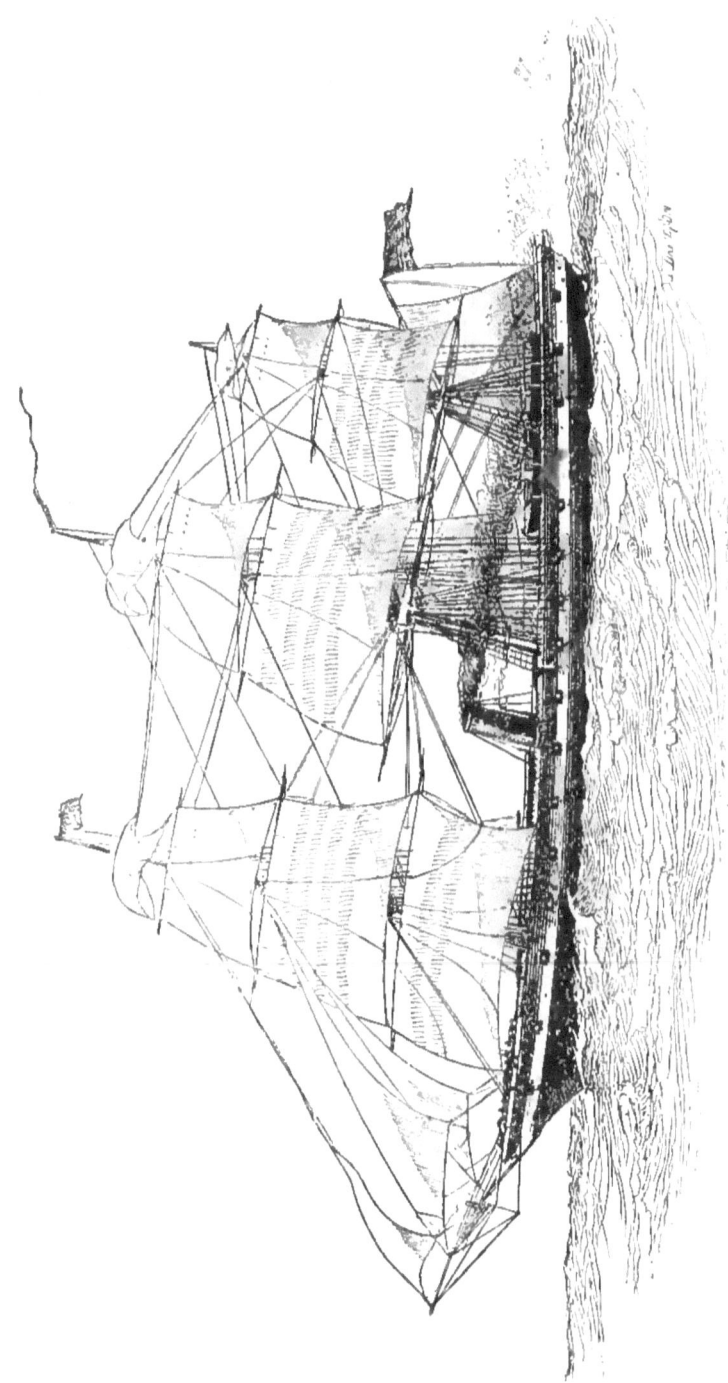

U. S. SLOOP-OF-WAR BROOKLYN

U.S. SLOOP-OF-WAR KEARSARGE

In 1859 a new type of sloop-of-war was introduced, of which the *Kearsarge* will serve as an example. This ship was built at Kittery, Maine; her length is 199 feet, beam 33 feet, draught of water 15.9 feet. She has a displacement of 1550 tons, and attains a speed of 11 knots per hour with an indicated horse-power of 812. The capacity of her coal-bunkers is 165 tons. Her battery consists of two 8-inch muzzle-loading rifles (converted), four 9-inch smooth-bores, and one 60-pounder. This has proved a very handy class of vessel, and for the year in which they were built were considered as having very fair speed under steam, the proportion of space occupied by boilers and engines being more than had been assigned in previous constructions. Several ships of this class were launched and put in commission before the war, and gave a new impetus to construction.

The types of vessels that were built during the war were selected for special purposes. The effort was made to multiply ships as rapidly as possible to blockade the coast and to enter shoal harbours; the "ninety-day gun-boats" and the "double-enders" were added to the navy list, and merchant steamers were purchased, and were armed with such batteries as their scantling would bear. All of these vessels have disappeared, with the exception of the *Tallapoosa*. The *Juniata and Ossipee*, of the *Kearsarge* type, but of greater displacement, were launched in 1862, and are still in service; and at about the close of the war several vessels of large displacement and great speed were launched which were never taken into service, have been disposed of since, and form no part of our present navy.

The *New Ironsides* and the *Monitor* represented the two features of construction which, produced in that period of emergency, have continued to impress naval architecture.

As a sea-going ironclad the *New Ironsides* was, for the time and service required, a success. She was built at the yard of Mr. Cramp, in Philadelphia, in 1862. Her length was 230 feet, beam 56 feet, draught of water 15 feet. She had a displacement of 4015 tons, and attained a speed of six knots per hour with an indicated horse-power of 700. The capacity of the coal-bunkers was 350 tons. Her battery consisted of twenty 11-inch smooth-bore guns. She was built of wood, and was covered with armour four inches in thickness, which, with the inclination given to her sides, made her impervious to the artillery that was used against her during the war. In one engagement with the batteries on Sullivan's Island, Charleston Harbour, lasting three hours, she was struck seventy times, but at the end of the action, except some damage to a port shutter or two, she with-

U. S. Iron-clad New Ironsides

U. S. Monitor Passaic

drew in as perfect fighting condition as when the fight commenced. This ship does not appear on the navy list, as she was destroyed by fire off the navy-yard at League Island, Pennsylvania.

The *Monitor* was, without doubt, the most remarkable production of the constructive art that appeared during the war. The original *Monitor* was lost at sea, but our illustration presents the *Passaic* class of Monitors, which quickly followed the original of this type.

The *Passaic* was built of iron, and was launched in 1862. Her length is 200 feet, beam 46 feet, draught of water 11.6 feet. She has a displacement of 1875 tons, and attains a speed of seven knots per hour with an indicated horse-power of 377. The capacity of her coal-bunkers is 140 tons. Her battery consists of one 15-inch smooth-bore and one 11-inch smooth-bore. Her sides are protected by five inches of laminated iron, and her turret by eleven inches of the same. This vessel and eleven others of her class constitute the entire armoured fleet of the United States. Too much credit cannot be awarded to Captain Ericsson for his brilliant conception of this floating battery, and the navy must be ever grateful to him for preserving it from the dire disaster which was averted by the appearance of the original *Monitor* at the moment of a great crisis. These vessels bore themselves well through the storms of elements and battle during the war, proving capable of making sea-voyages, and of resisting the effects of the artillery that was in use during the period of their usefulness; but an interval of more than twenty years has produced such a change in artillery

U.S. DOUBLE-TURRETED MONITOR TERROR

as to make the protection afforded by a few laminated plates of one-inch iron but a poor defence against weapons which have robbed this fleet of its once formidable character. Although many of the features of the original design may be retained in new constructions, most of the details will be changed, notably in the turret, in consequence of the greater weight resulting from the increased thickness of armour. The central spindle around which the Ericsson turret revolves must disappear, and the turret must turn on rollers under the base.

The effect produced abroad by the success of Ericsson's *Monitor* is so familiar to all that it hardly needs more than a passing allusion here. There is no doubt that the *Monitor* was the progenitor of all the turreted vessels in the fleets of the world, though the essential principle of the vessel, however, was never viewed with favour. This principle consists in the low freeboard, which, besides reducing the size of the target, is intended to contribute to the steadiness of the hull as a gun platform by offering no resistance to the waves that are expected to wash freely over the vessel's deck: the horizontal overhang of the *Passaic* class is intended to contribute to resisting a rolling motion. The vessel was designed to be as a raft on the water, constantly submerged by the passing weaves, hermetically sealed to prevent the admission of water, and artificially ventilated by means of blowers drawing air down through the turret. This was the most startling feature about the construction. The protection afforded to the battery by a circular turret having the form best suited to deflect projectiles, the employment of machinery to point the guns by the rotation of the turret, the protection to motive-power, to anchoring apparatus, etc., all presented admirable points of advantage, but the almost perfect immersion of the hull, and the absence of motion due to the *great stability*, are the essential features in the construction.

The double-turreted Monitors, of which the *Terror* indicates the class, were built with a sponson, and it would have been better for the navy if this had been the only deviation made from the original design of Captain Ericsson. But it was not; the great mistake was made of building this class of Monitors of wood—a style of construction which had been already condemned abroad, in consequence of the impossibility of repairing an armoured vessel so constructed, it being necessary to re-move the armour for that purpose.

The *Miantonomoh, Monadnock*, and *Terror* were completed and put in commission. The *Miantonomoh* made a cruise to European waters, spreading the fame of Ericsson, and proving the ability of a vessel of this type to navigate the high seas; the *Monadnock* made the voyage to

U. S. Frigate Tennessee

the Pacific, passing through the Strait of Magellan; and the *Terror* was for a time on service on our eastern coast; but their lifetime was of short duration, and they are now being rebuilt, or rather new vessels, three of which bear their names, are now under construction of iron, which will serve to make them efficient and durable.

It will hardly be a digression at this point to call attention more particularly to these double-turreted Monitors now under construction. They bear the following names, *viz.*, *Puritan, Terror, Amphitrite, Miantonomoh,* and *Monadnock.* There was much contention about the completion of these vessels, and imaginary defects were ventilated in the newspapers. It may be that these attacks and erroneous statements prejudiced the public mind, and that the idea was entertained by some whose opinion is valued that there were grounds for the doubts that had been expressed of their sea-worthiness. The practical effect of these statements was to prevent Congress from appropriating money for the completion of the vessels, and this course on the part of Congress might have confirmed some in their doubts. Several boards of officers, most competent experts, however, reported upon them, recommending their completion; of these that made by the Advisory Board may be regarded as a final decision, for it was accepted without question. The Advisory Board reported as follows:

> It is our opinion that it would be wise and expedient to finish these vessels at once, and for the following reasons, *viz.*:
>
> 1. The hulls, as they are at present, are of excellent workmanship, fully up to the present standard condition of *iron* ship construction, whilst the flotation of the *Puritan* and the behaviour of the *Miantonomoh* at sea confirm the correctness of the calculations of the designs.
>
> 2. It is easily possible to complete the vessels by taking advantage of the recent developments in armour, guns, and machinery, without making any radical changes in the designs, so that their speed, endurance, battery power, protection, and sea-going qualities shall be fully equal to those of any foreign ironclad of similar dimensions designed previous to 1879.
>
> 3. The vessels may be finished so as to develop all the above-mentioned advantages without making their total cost, when completed, in any way exorbitant, compared with the results obtained; again, the interests of our sea-coast defence require a force at least equal to that which would be represented by these vessels.

U. S. sloop-of-war Adams

We take the liberty of calling your attention to a certain erroneous impression which now exists with regard to these vessels. In one of the reports of these hulls a doubt was thrown on the correctness of the calculations of the *Puritan*. This doubt has spread in the public mind until it includes all the ships. The actual flotation of the *Puritan* and the *Miantonomoh* proves beyond question not only the reliability of the calculations, but also that the hulls of these vessels are lighter in proportion to the total displacement than those of any ironclad low freeboard hulls afloat, with two exceptions.

It has been the unfortunate custom, in arguments as to the value of the results to be obtained, to compare these vessels with such foreign ships as the *Inflexible* and the *Duilio*, to the evident disadvantage of the Monitors, no account whatever being taken of the fact that these vessels are double the size of the Monitors. If these hulls be compared with foreign ones of similar dimensions, no such disparity will appear.

These vessels, with the exception of the *Monadnock*, have their machinery in place; the *Miantonomoh* has her side armour on; the others are finished as to their hulls, except the interior fittings, side armour, and turrets. The estimated cost to complete them is about four millions of dollars. When we consider the very slight defence that the country now possesses in the single-turreted Monitors before alluded to, it would seem imperative to complete with all despatch these vessels, which would represent a force of real power.

These vessels are of iron as to the hulls, but they will be armoured with steel or compound armour, and will be armed with the most powerful modern artillery that can be accommodated in their turrets. Their names appear in the navy list as "building." They were launched in 1883.

The double-decked ship *Tennessee* was the only frigate, or "first-rate," borne, up to within a few months, on the list of vessels of the navy as available for sea-service. She was for many years in commission as the flag-ship of the North Atlantic Station, but this year she reached that condition when the twenty *per cent*, law consigned her to "ordinary," from which she has lately been removed under the operation of the hammer of the auctioneer. She was launched in 1865. Her length was 335 feet, beam 45 feet, draught of water 21.8 feet. She had a displacement of 4840 tons, and attained a speed of 11 knots with an indicated horse-power of 1900. The capacity of her coal-bunkers was

U. S. SLOOP-OF-WAR MARION

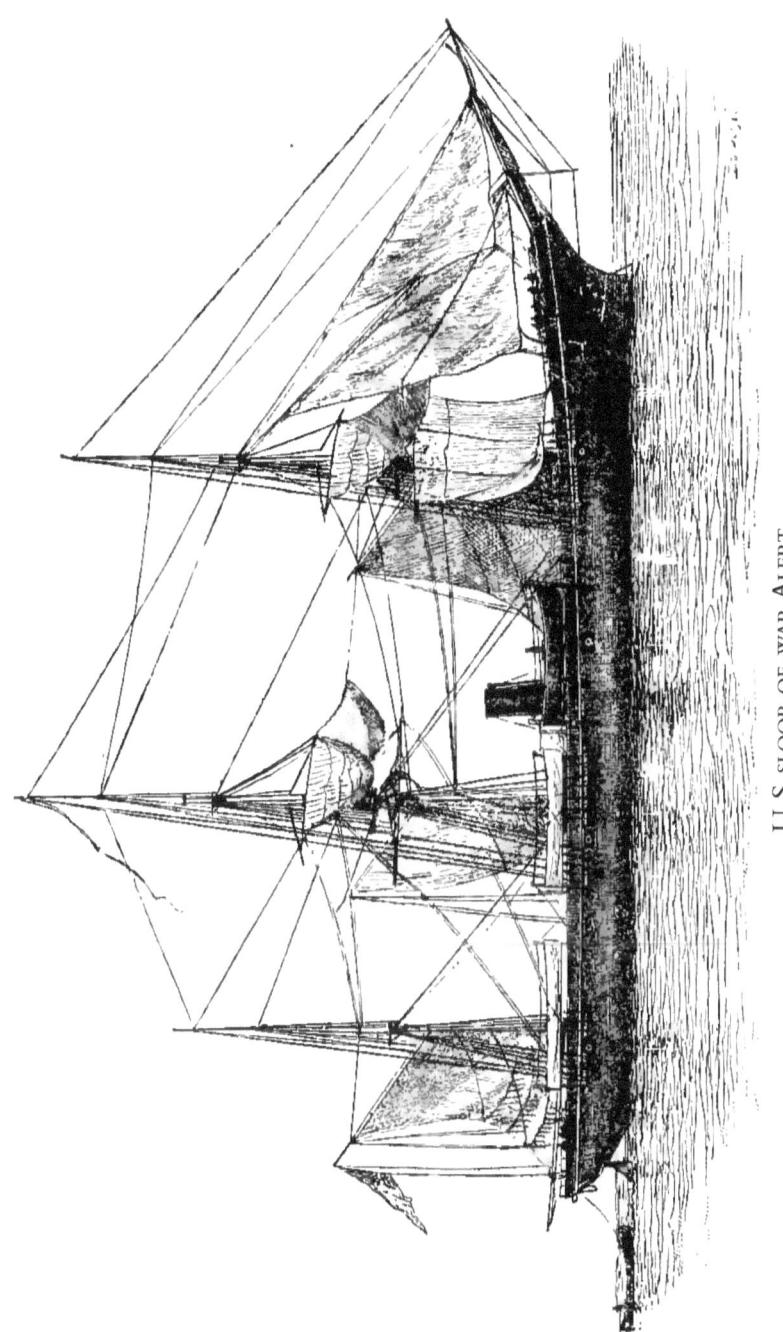

U. S. SLOOP-OF-WAR ALERT

381 tons. Her battery consisted of two 8-inch muzzle-loading rifles (converted), sixteen 9-inch smooth-bores, and four 80-pounders.

The vessels next in order of construction are those of the *Adams* class, small sloops-of-war, which were launched in 1874.

These vessels were built of wood. They are convenient and handy, and perform the duty required of a cruiser in time of peace. Engine-power is developed in them to a higher degree than in those preceding them, but in all else they are merely a repetition of earlier constructions. The *Adams* was launched in 1874. Her length is 185 feet, beam 35 feet, draught of water 14.3 feet. She has a displacement of 1375 tons, and attains a speed of 11.3 knots with an indicated horse-power of 715. The capacity of her coal-bunkers is 140 tons. Her battery consists of one 11-inch and four 9-inch smooth-bores, and one 60-pounder.

The *Marion* class of sloops, launched about the same period, are of an increased displacement and speed, and built of wood. The length of the *Marion* is 216 feet, beam 37 feet, draught of water 16.6 feet. She has a displacement of 1900 tons, and attains a speed of 12.9 knots per hour with an indicated horse-power of 966. The capacity of her coal-bunkers is 135 tons. Her battery consists of one 8-inch muzzle-loading rifle (converted), six 9-inch smooth-bores, and one 60-pounder.

The *Alert* is one of three vessels that were built of iron in 1874, the exceptional and spasmodic indication of an effort to change the material for construction, much induced by pressure from the iron interests of the country. This effort was made in a very mild and tentative manner, and was limited to this small class of diminutive vessels. The length of the *Alert* is 175 feet, beam 32 feet, draught of water 12.9 feet. She has a displacement of 1020 tons, and attains a speed of ten knots per hour with an indicated horse-power of 655. The capacity of her coal-bunkers is 133 tons. Her battery consists of one 11-inch and four 9-inch smooth-bores, and one 60-pounder.

The shock attending the first step towards a change in the material for construction was so great as to cause a suspension of the effort, and in 1876 was launched the *Trenton*, built of wood, which represents the latest of that type on the list of the navy. The length of this ship is 253 feet, beam 48 feet, draught of water 20.6 feet. She has a displacement of 3900 tons, and attains a speed of 12.8 knots per hour with an indicated horse-power of 2813. The capacity of her coal-bunkers is 350 tons. Her battery consists of ten 8-inch muzzle-loading rifles (converted).

The above is a fair presentation of our old navy. Of such vessels we have, larger and smaller, twenty-five which are fit for service as cruisers, exclusive of the old single-turreted Monitors. These cruisers are built of wood, have low speed, and are armed with smooth-bore guns, with a sprinkling of rifled cannon, converted on the Palliser system from smooth-bore cast-iron guns. Of what service is this force, this relic of a past age?

The duties of a navy, apart from the consideration of war, are manifold. As stated by the first Advisory Board, it is required for:

> Surveying, deep-sea sounding, the advancement and protection of American commerce, exploration, the protection of American life and property endangered by wars between foreign countries, and service in support of American policy in matters where foreign governments are concerned.

With such a poor force it must be evident that it was impossible to discharge in an efficient manner all the duties of a navy. Our work in foreign surveys is limited to that of one small vessel on the west coast of North America; our deep-sea soundings are few and far between, dotted along the tracks pursued by our ships while going to and returning from distant stations; our commerce is protected; but we are unable to support any positive policy that the government might decide to declare in reference to, for example, the Monroe doctrine. To say nothing of European naval armaments, it is only necessary to point to some of the smaller powers in our own hemisphere that possess ships-of-war with which we have nothing fit to cope.

Our people cannot desire to assume a position in the society of naval powers without supporting the position with dignity; they cannot wish their navy to be cited as a standard of inefficiency; they cannot wish to force their representatives (the officers of the navy) into a position of humiliation and mortification such as is imposed by being called on to deprecate criticism by laboured explanations. Better abolish the navy and lower our pretensions.

But the fact seems to be that the rapidity of naval development has not been properly appreciated, and it is after a long interval of indifference that, attention being at last centred on the subject, it is seen how rapid its strides have been, and how utterly we are distanced in the race. There is evidently now in the country a growing desire to repair the effects of the past oversight, and we see Congress has moved in the matter. As all political parties now unite in the necessity of effort in this direction, the hope is inspired that the subject is to be separated

U. S. SLOOP-OF WAR TRENTON

from those of a partisan character, and that the rehabilitation of the navy will be put on its proper level, and accepted as a national question in which all are alike interested.

Possessed as we are now of a navy such as has been indicated, the change that was instituted involved a most violent transition. In reviewing our work of construction for over thirty years we saw no new type of cruiser. The only types of ships that we produced were those that date before the war; since which we but reproduced the same in classes of differing dimensions. From the sailing-ship with auxiliary steam-power we passed to the steamer with auxiliary sail-power; but we had no full-powered steamers, with or without sails. As long as it was considered necessary to spread as much canvas as was used, the space assigned to boilers and engines was limited, and we failed to achieve full power; and a reduction to the minimum of sail-power had to be accepted before we could present a type of a full-powered steamer.

With the exception of two vessels of the *Alert* class built of iron, we had nothing but wooden hulls. We had continued to build in perishable material, requiring large sums to be spent in repairs, and ignoring the manufactures of the country which could have been aided in their development by the contrary course. We permitted the age of steel to reach its zenith without indicating that we were aware of its presence.

In these ships, with the exception of a few converted rifles of 8-inch calibre, our armaments consist of smooth-bore cast-iron guns which have composed our batteries for thirty years. These are now to be discarded, and their places to be filled with modern steel cannons.

Torpedoes, movable torpedoes, of which we know nothing practically, are to be brought to the front, and are to form part of our equipment. Torpedo-boats are to be brought into use, and details innumerable are now to be studied and worked out.

Conceive, then, a high-powered steamer with a minimum of canvas, built of steel, armed with modern steel artillery, and a secondary battery of Hotchkiss guns, fitted for launching movable torpedoes, with protective deck over boilers and engines, divided into many watertight compartments giving protection to buoyancy, and compare such a ship with the old type of the United States cruiser, and an idea may be formed of the violence of the transition through which we had to pass. And there was nothing intermediate to break the suddenness of this change; there was no connecting link. The structure of today was placed in direct contrast with that of twenty-five years ago, (as at time of first publication). This is the position in which we stood, and we could only accept a situation from which there was no escape.

From all appearances the navy is now to be given an opportunity of asserting itself, and the steps already taken to remedy the existing state of things can be stated in a few words.

The origin of the effort dates from June, 1881, when the first Advisory Board was appointed to consider and to report on the need of appropriate vessels for the navy. This Board, in its report of November 7, 1881, decided that the United States navy should consist of seventy unarmoured cruisers of steel; it reported that there were thirty-two vessels in the navy fit for service as cruisers, and it indicated the character of the new vessels to be built. This Board confined itself to the consideration of unarmoured vessels, as it did not consider that the orders under which it acted required that it should discuss the subject of armoured ships, though it expressed the opinion that such vessels were indispensable in time of war.

Some time elapsed before any practical results followed from the action of this Board, but in an act of Congress approved March 3, 1883, the construction of three steam-cruisers and a despatch-boat was authorized. These vessels, the *Chicago, Boston, Atlanta,* and *Dolphin,* are, with the exception of the *Chicago,* now in commission.

In an act of Congress approved March, 1885, five additional vessels were authorized, and these, the *Charleston, Baltimore, Newark,* and gunboats No. 1 and 2, are under construction.

Up to the time of the inception of these cruisers no steel for ship-plates had been rolled in the United States. Construction in American iron plates had been extensively carried on, but if steel plating was required it had to be imported at great cost to the builder. Those who contemplated bidding on the proposals issued by the government for the first four vessels had to consider this matter. Mr. John Roach, of New York and of Chester, Pennsylvania, undertook the manufacture of this material, and finding that success attended his experiment, he was able to direct extensively the steel-works at Thurlow, Pennsylvania, to this line of business, and when the bids were opened it was found that this new industry, introduced through his enterprise, enabled him to underbid all competitors. After receiving the contracts for the ships, Mr. Roach contracted with the Phoenix Iron Company, of Phoenixville, with Messrs. Park Brothers, of Pittsburgh, and the Norway Iron and Steel Works, of South Boston, for supplies of similar material: thus the first step in this effort to rehabilitate the navy resulted in introducing a new industry into the country. The still more extensive development of industries that will attend the work of rehabilitation as it advances will be treated further on.

U.S. Frigate Chicago

Before presenting the types of cruisers which are now to be introduced into the navy, it may be well to refer to an error that exists, or has existed, in the popular mind as to the signification of a steel cruiser. To many who are uninformed in technical language the word steel, in connection with a vessel of war, implies protecting armour, and such misapprehension would convey the idea that a cruiser of steel is able to contend with an armoured vessel. This is a mistake; there is protection obtained by constructing a vessel of steel, but not such as is provided by armour. The destructive effect of shell-firing and the development in modern artillery have made armour necessary for all vessels which can carry it, and has also made it necessary to provide all other protection possible to vessels that cannot carry armour. Although this protection cannot be given absolutely to the hull of such ships and to the *personnel*, it is provided to the *buoyancy* by the introduction of water-tight compartments and protective decks, which limit the destructive effect of the fire of the enemy and localize the water that may enter through shot-holes. With a wooden hull it would not be possible to combine this precaution because of the difficulty in making joints water-tight between wood and metal, and in consequence of the weight that would be added to a wooden hull, which is already from sixteen per cent, to twenty per cent, heavier than if constructed of steel. The only defensive advantage possessed by a steel unarmoured cruiser over a wooden one is derived from this system of construction.

The *Chicago* is a steam-frigate, built throughout of steel of domestic manufacture, the outside plating being $9/16$ inch thick. Her length is 325 feet, beam 48.2 feet, draught of water 19 feet. She has a displacement of 4500 tons, and will attain a minimum speed of 14 knots per hour with an indicated horse-power of 5000. The capacity of her coal-bunkers is 940 tons, and she carries a battery of four 8-inch steel breech-loading guns in half-turrets, and eight 6-inch and two 5-inch steel breech-loaders on the gun-deck. This ship has nine athwartships bulkheads, dividing the hull into ten main water-tight compartments, and the machinery and boilers are covered by a protective deck one and a half inches in thickness. When the bunkers are full of coal she has a coal protection nine feet thick from the water-line to eight feet above it.

The deck plans show the arrangement of the main battery, in addition to which she carries a powerful secondary battery of Hotchkiss rapid-firing single-shot, and revolving cannons and Gatling guns.

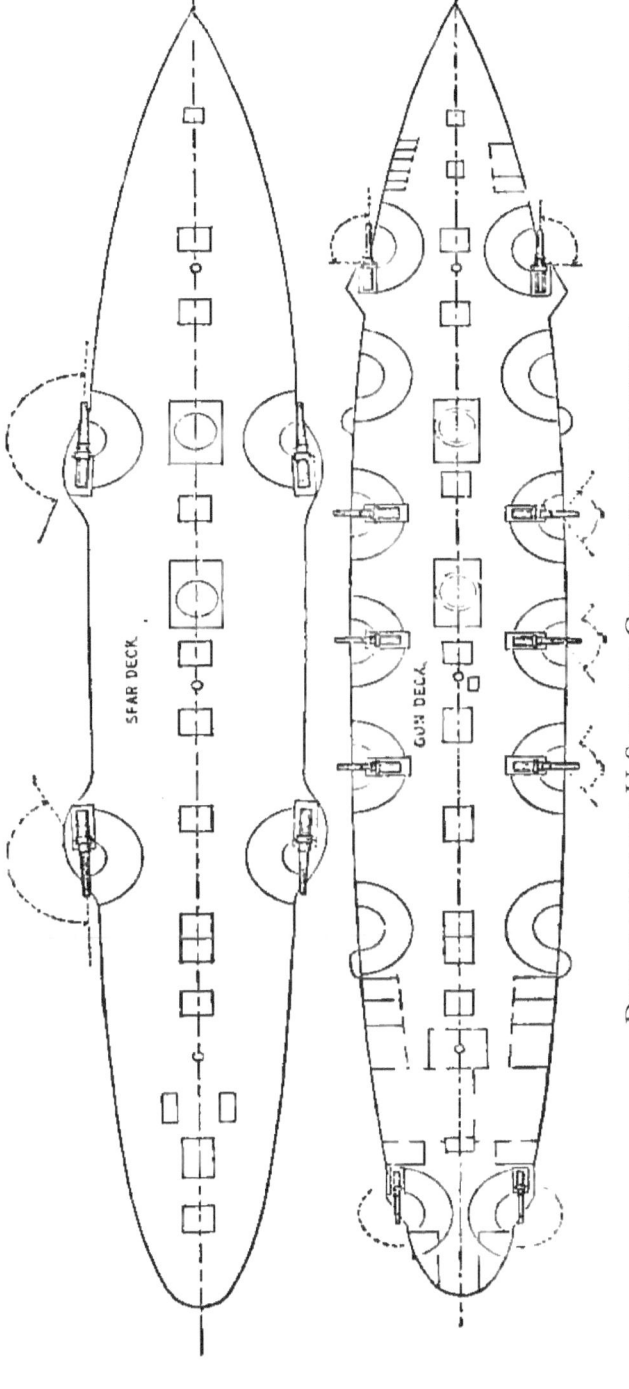

Deck plans of the U. S. frigate Chicago, showing battery

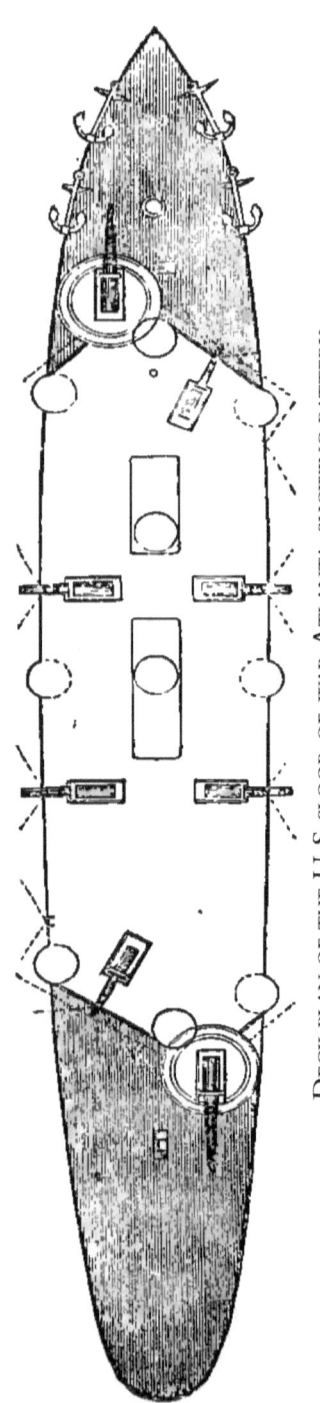

DECK PLAN OF THE U. S. SLOOP-OF-WAR ATLANTA, SHOWING BATTERY

The bow of the vessel is strengthened for using the ram with which she is fitted. The rudder and steering-gear are under water. She has two screws—a subdivision of power which is given to all ships-of-war of over 3000 tons displacement—from which a great advantage is derived if one engine is broken down, as three-fourths of the speed can be maintained with the other. The advantage of this in a naval action is obvious.

The *Atlanta*, of which the *Boston* is a counterpart, presents another type of a steel unarmoured cruiser. She is a steam-corvette, or sloop-of-war, a single-decked ship. Her length is 276 feet, beam 42 feet, draught of water 16 feet 10 inches. She has a displacement of 3000 tons, and has attained a speed of 15.5 knots per hour with a maximum horse-power of 3482. The capacity of her coal-bunkers is 580 tons, and her battery consists of two 8-inch steel breech-loading guns and six 6-inch, besides a secondary battery of Hotchkiss and Gatling guns.

In vessels of this class it is usual to have an open-deck battery, with a poop-deck and top-gallant forecastle at the extremities, but the effort has been made in this type to increase the effectiveness of the battery by giving the guns a more extended lateral train than is possible when a ship is arranged with a forecastle and poop-deck. These, with the accommodations which they provide, have been removed from the ends of the ship, and a superstructure has been erected amidships. This arrangement gives a clear sweep forward and aft for the powerful 8-inch guns, enabling the forward gun to cover an all-around fire of from 40° abaft the beam on the port side to 30° abaft the beam on the starboard side, the after 8-inch gun having a corresponding lateral sweep aft. Within the superstructure are mounted the six 6-inch guns, two on each side on the broadside, with a train of 60° before and abaft the beam, the other two being mounted at diagonally opposite corners in such a way as to admit of their use either on the broadside or for fire ahead or astern. This object is achieved by mounting the two 8-inch guns *en échelon*, the forward gun being on the port side of the centre line of the ship, and the after 8-inch gun on the starboard side of the same line. This is shown on the deck plan.

It does not require the discrimination of a professional eye to see the increased power given to the battery by this arrangement. It is an innovation that was very startling to the conservative mind; but the more familiar the idea becomes, the more favourable opinion grows to the change, and the more apparent becomes the increased offensive power of the ship. The extremities of this type of ship will not, of course, be so dry in heavy weather as if it had a forecastle and poop, but it must be remembered that every part of the spar-deck is from nine to ten feet above the water. The rig of the *Atlanta* is that of a brig, but without head-booms; the fire ahead of the forward guns is thus unobstructed, and the ram with which she is fitted is always clear for use. The division of the hull into water-tight compartments by athwartships bulkheads, and a protective deck over engines and boilers, form a part of the construction.

The *Dolphin*, though not regarded as a vessel for fighting purposes, is the type of a class that is needed in all navies for duty as a despatch-boat, or for the temporary accommodation of a commander-in-chief of a squadron who may desire to communicate rapidly with his ships at distant points. She is well fitted for the service, and is now in commission, demonstrating her ability to perform the work required of her. She could also be of service as a commerce destroyer, for which service she is equipped with one gun of long range. Her length is 240 feet, beam 32 feet, draught of water 14.25 feet. She has a displacement of 1485 tons, and attains a speed of 15 knots per hour.

Her advent into the navy marks an epoch—the inauguration of the successful manufacture in the United States of American rolled steel ship-plating, equal to the best in the world, as shown by the most rigid government tests. The *Dolphin* is the first vessel, whether for naval or commercial purposes, that is built entirely of steel of domestic manufacture, and is the pioneer representative of other similar

The Atlanta

industries which will be developed as the rehabilitation of the navy proceeds. She has proved herself eminently successful, and is the fastest sea-steamer of her displacement built in the United States, with perhaps the exception of the steam-yacht *Atalanta*. She is a stanch vessel of great structural strength, and does credit to the ship-building profession of the country.

Of the additional cruisers authorized by the late acts of Congress, particulars will be found in the Notes.

In one of the larger vessels the type of the *Atlanta* will be reproduced on a larger scale, while the other vessel of the same class will be provided with a poop and top-gallant forecastle, and will carry her forward and after guns on sponsons, by which means fire ahead and astern will be secured. This will make it necessary to limit the power of the battery of the second vessel to 6-inch guns, as the 8-inch gun cannot conveniently be carried on sponsons in a vessel of 4000 tons displacement.

The heavy gun-boat will carry six 6-inch guns, the forward and after ones on sponsons; and the light gun-boat will carry four of the same guns. In the construction of these additional vessels advantage has been taken of all our experience in our initial effort, and of whatever developments may have been made by others since the earlier vessels were designed. The absolute departure from the old standards is apparent in material, in armament, in speed, and in rig.

The causes that have led to this change in material may be found, first, in the change that has taken place in ordnance. The introduction of the rifled cannon, and its subsequent development, have increased very much the weight of this part of the equipment of a vessel-of-war, and the necessity of accommodating the stowage of charges of powder much increased in size, and of ammunition for the secondary batteries, which must be most liberally supplied, makes an absolute demand on an increased portion of space. Again, the increased speed now considered indispensable makes a similar demand for space, and carries with it as well an increased proportion of the total displacement. In a wooden hull it would be impossible to reconcile these demands, in consequence of the weight of the hull itself.

The hull and hull fittings of an unarmoured cruiser built of wood will weigh from 49 *per cent,* to 52 *per cent,* of the total displacement. With high-powered engines it is doubtful if sufficient strength can be obtained with even 52 *per cent,* of the displacement for the hull, and this must suppose the absence of all protection to buoyancy, as water-tight compartments.

The hull and fittings of a steel. cruiser, exclusive of protective decks, will weigh from 39 per cent, to 44 *per cent*, of the total displacement.

Suppose a 4500-ton ship built of wood weighing 50 *per cent*, of the total displacement, and the same ship built of steel weighing 40 *per cent*, of the total displacement, the respective weights of the hulls will be 2250 tons and 1800 tons, a difference of 450 tons, the steel hull being one-fifth, or 20 *per cent.*, lighter than the wooden one. This will allow for increased weight of ordnance, protective deck, or increased coal endurance, as may be decided when considering the service on which the ship is to be employed.

But notwithstanding the saving thus obtained, the question of weights is still full of difficulties and embarrassments, and it is found impossible in the same structure to accommodate all demands from the different departments concerned in the equipment of a vessel-of-war. The sail-power has been reduced, so as to save weight of spars and sails, which have become of secondary importance, but this will not satisfy all the requirements of the problem. As articles appertaining to the old method of equipment are removed, those belonging to what are considered necessary under the new order of things are brought forward. Space is still to be found for movable torpedoes, for torpedo-boats, and for engines and appurtenances for electrical apparatus for lighting the ship, for searchlight, and other ordnance purposes. It is evident that much study is needed to reduce weights in all the essential parts, so as to be able to accommodate all the devices which the progress of ideas continues to present. Much is yet to be done by the substitution of steel for iron in many parts of our engines, and experiments abroad lead to the hope that the weights of boilers may be much reduced, but as the question stands today it is impossible to provide any single ship with all the appliances that are considered necessary for a perfectly equipped vessel-of-war. Every ship, therefore, must present a compromise.

Another reason for the transition from wood to steel hulls is the durability of steel as compared with wood. Referring to the large sums of money that have been appropriated under the head of construction and repairs, for which there is now so little to show (and disregarding the question of administration, which of course is vital, but which has no place in this chapter), the main reason for the deficiency in the results is that all this money has been expended in perishable material. Every ship that has been built of wood since the war has been a mistake. The most serious error was committed when the wooden double-turreted Monitors of the *Miantonomoh* class were built, which,

U. S. DESPATCH-BOAT DOLPHIN

it is believed, was done against the protest of Captain Ericsson, The result was the early decay of these vessels, and the present defenceless condition of our sea-coast. The lifetime of a wooden ship is of short duration. It requires constant repairs, which amount in the long-run to rebuilding, and it is in this manner that so many of our old ships are still retained in service, (as at time of first publication); but in the case of a wooden armoured vessel these repairs are impossible without removing the armour. This was the condition of affairs with regard to these Monitors, and the consequence is that the country has to incur the expense of entirely new constructions. These are in durable material, and will give good account of themselves when called on.

The steel hulls that it is now the intention shall compose the fleet, will, if well cared for, endure in perfect condition for thirty years. In fact, the lifetime of an iron or steel hull is not defined to any limit, and if a perfect anti-corrosive and anti-fouling composition can be produced, the limit may be regarded as indefinite.

The foregoing remarks on our new navy apply to unarmoured cruisers, a class of ships which supply a want in time of peace, but can-not fulfil the purposes of war. At such a time the armoured ship is recognized as indispensable, and there is every reason why the construction of armoured vessels should proceed simultaneously with that of the unarmoured cruisers. These are a more intricate problem for study, need much more time to build, and are required, while at peace, as a school of instruction in which to prepare for war. Our selection of armour has been much assisted by the investigations of others, and we are in a favourable condition to make a decision on this point; and the type of vessel best suited for a cruiser seems to be settled, by the uniform practice of foreign nations, in favour of the barbette.

It must be remembered, however, that six months since we were not in a condition to proceed with the construction of armoured vessels, depending on our own resources. We had to go abroad to purchase armour, or set ourselves to the task of establishing works where it could be manufactured. The establishment of these plants was the first thing needful, and until this was done it was impossible to make ourselves independent in this matter. The construction of our first unarmoured cruisers introduced into the country the industry of rolling steel ship-plates; the construction of our new ordnance and armoured ships has, in turn, introduced the new industries of casting and forging large masses of war material.

This subject, so far as it relates to ordnance, was referred to a mixed board of army and navy officers, known as the gun foundry board;

this, with the aid and counsel of some of the ablest and leading steel manufacturers in the United States, submitted to Congress a report which presented a solution of the problem, and demonstrated on what terms the steel manufacturers of the country could be induced to work in accord with the government. The Board had under consideration only the subject of foundries and factories for gun construction, but the casting and forging facilities required for guns could be applicable to armour; thus in providing means for the manufacture of one, the other purpose was equally subserved.

With material of domestic manufacture at hand, it will be the duty of the government to provide the navy with a fleet of ten armoured cruisers of the most approved type. These vessels would form the outer line of defence of the coast during war, and should be of such force as to be able to contend with any second-class armoured vessel of other nations. Some of them should be always in commission during times of peace, if only for instruction and practice purposes, and one should be assigned to each squadron abroad to carry the flag of the rear-admiral in command, to assert our position in the society of naval powers, able to give substantial "support to American policy in matters where foreign governments are concerned."

The ability to contend with armoured vessels of the first class must be reserved for another type of ships, which are styled "coast-defence vessels," and without which our new navy will not be thoroughly equipped for contributing its full share to defence at home. In considering armoured vessels, what was said before as to the character of compromise that obtains in vessels-of-war must be borne in mind. All desirable features cannot be concentrated in any one ship; the special duty for which the vessel is to be used controls the selection. The sea-going armoured cruiser is expected to keep the sea for a lengthened period: she must have large coal endurance. She may be called on to sustain more than one engagement: her supply of ammunition must be large. Her speed must equal that of the fastest sea-going vessels of similar type to enable her to pursue an equal or to avoid a superior force: hence much space and displacement must be assigned to engines and boilers. Thus the amount of her armour and the weight of her battery are affected by these other demands, which are the more imperative.

In the case of coast-defence vessels the conditions are changed, enabling in them the full development of both offensive and defensive properties. These vessels are assigned to duty on the coast: they must be as fit to keep the sea as are the armoured cruisers, and they must

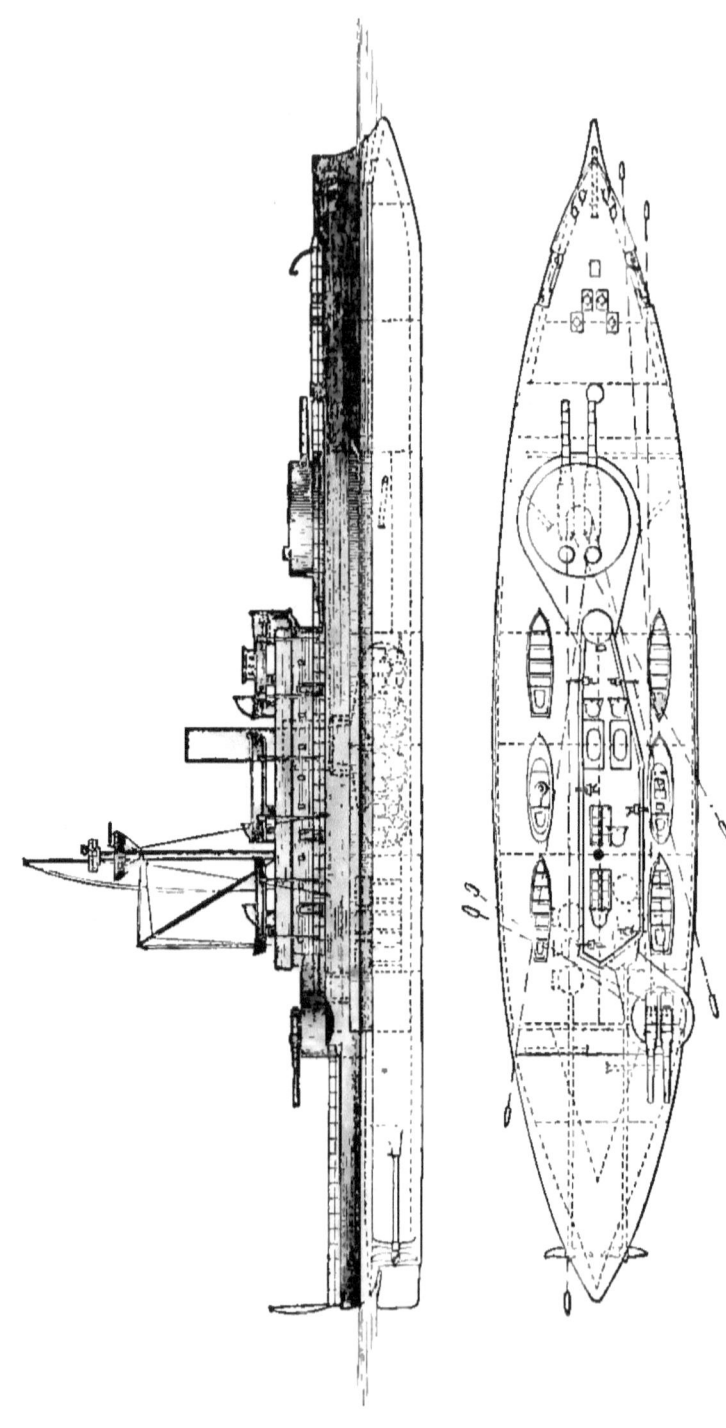

Light draught coast-defence vessel, with deck plan

be able to fight their guns in all weathers that the armoured cruiser can fight; but as they do not require the coal endurance nor the speed of the ship that is to keep the sea for lengthened periods, the weight saved in coal and machinery can be utilized in battery and armour. Such vessels constitute the main line of naval defence, as they can be made almost absolutely invulnerable and irresistible.

Under an act of a late Congress a board on "fortifications and other defences" was occupied in considering the defences of the coast, and there were recommended by this Board two classes of "floating batteries" (so called), coast-defence vessels, and one class of low free-board vessels for harbour defence. An examination of the designs of these vessels shows that they are replete with merit, and present some novel and valuable features. A justifiable limitation is put on the coal endurance and speed, though fair speed is secured; and altogether the plans prove there can be designed vessels of comparatively small dimensions, light draught, great handiness and manoeuvring power, which can carry the heaviest guns, and be capable of contending on equal terms with the heaviest European battleships. The cut below represents the smaller of the type of coast-defence vessels.

The largest class will be armed with two 107-ton guns in a turret, and two 26-ton guns in a barbette. The thickness of armour will vary from 16 to 18 inches.

The second class will be armed with two 75-ton guns in a turret, and two 26-ton guns in a barbette. The thickness of armour will be from 11 to 16 inches.

The smallest vessels, for harbour defence, modified Monitors, will be armed with two 41:-ton guns in a turret, and two 26-ton guns in a barbette. The thickness of armour will be from 10 to 13 inches.

A fleet composed of such vessels as are represented in the largest type would be able to engage an enemy at some distance from the coast—an important object in these days when the range of heavy rifled cannons makes it possible to shell towns from a great distance, and at points remote from shore batteries.

Nominally we have now a fleet of vessels for coast defence, the old war Monitors of the *Passaic* class; but the contrast between them and the vessels recommended by the Fortifications Board is about equal in degree with that between our wooden fleet and the new steel cruisers.

It is intended that a movable automatic torpedo shall be utilized by all armoured vessels, either by means of a torpedo-boat to be carried by armoured cruisers, or by the vessel itself in the case of coast and harbour defence ships.

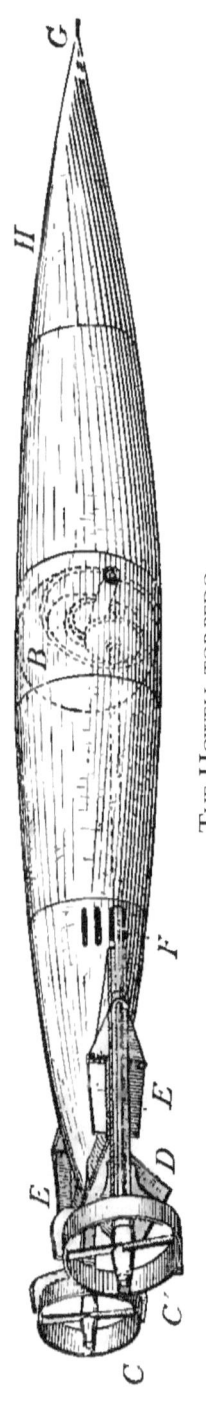

THE HOWELL TORPEDO. *B*, fly-wheel. *C, C*, screw propellers. *D*, diving rudder. *E, E*, steering rudders. *F*, water-chamber containing automatic apparatus. *G*, firing pin. *H*, position of gun-cotton magazine.

The torpedo that has mainly succeeded thus far in recommending itself to the naval powers is that invented by Mr. Whitehead. Numerous efforts have been made by others in this field, but the difficulties that surround it are made very apparent by the paucity of the results. It will be understood that the torpedo, when launched, is left entirely to automatic control; hence, apart from the motive power, it is necessary that it shall possess directive power, vertically to control immersion and horizontally to control direction in the horizontal plane. In the Whitehead torpedo the immersion is well regulated, and if no deflecting influences are encountered, the direction is also preserved; but it fails where deflecting influences intervene. During the Turko-Russian war valuable experience was gained, and instances are known where the torpedo failed to operate from want of directive power. An instance is cited where a torpedo was deflected by striking the chain of a vessel at anchor, causing it to pass harmlessly to one side. Another instance is cited where the torpedo was deflected from the side of a ship owing to the angle at which it struck. It is evident that perfection cannot be associated with a weapon of this class that has not a strong directive force inherent in it.

The torpedo invented by Captain J. A. Howell, of the United States navy, possesses this property to an eminent degree, and it is regarded by most competent experts as the successful rival of the Whitehead. In the Howell torpedo the power is stored in a fly-wheel revolving with great rapidity in a longitudinal vertical plane, and its gyroscopic tendency makes it impossible for the torpedo to deviate from its original course in a horizontal plane; the principle is the same as insures the accuracy of the rifle-bullet, enabling it to resist deflecting influences. The latest experiments of Captain Howell in controlling the immersion of his torpedo were very successful, and it is probable that the automobile torpedo for our new navy will be an American invention. Liberality in experiments is indispensable in perfecting a device of this kind; it is to be hoped that such may be extended to the Howell torpedo.

The general reader is probably not aware of the effect on naval warfare produced by the introduction of the auto-mobile torpedo, affecting the constituents of the fleet itself. Formerly a fleet consisted of battleships alone, or with store-ships to provide consumable articles; to these were later added despatch-boats for the service indicated by their title; but since the introduction of the torpedo an additional fleet of torpedo-boats is considered necessary for the protection of the battleships. All armoured ships are expected to carry at least one torpedo-boat, which is designed for operating against the enemy during an action at sea, and the universal adoption of this practice has led to the introduction into fleets of a new type of vessel called torpedo-boat catchers, whose primary duty it is to destroy the torpedo-boats of the enemy. For this purpose these vessels have phenomenal speed, and besides their equipment of automobile torpedoes, are provided with powerful batteries of single-shot and revolving Hotchkiss guns, capable of penetrating all parts of a torpedo-boat. This type of vessel is now being tested by the English and the Continental governments, and forms one of the constituents of their fleets.

The torpedo-boat is undoubtedly one of the features that should be introduced into our new navy, not only for their possible use on the high-seas, but for the purpose of supplementing the harbour-defence vessels, while the type of vessel known as the torpedo-boat catcher would be a powerful auxiliary to the armoured cruisers on the first line, or the more powerful vessels forming the second line of the coast defence.

Notes

For the new navy of the United States Congress has authorized the construction of twenty-five vessels, of which seven will be armoured, sixteen unarmoured, and two "such floating batteries, rams, or other naval structures for coast defence" as may be determined by the Navy Department. This list embraces five double-turreted Monitors, one armoured battleship, one armoured cruiser, eight partially protected cruisers, one dynamite-gun cruiser, four gun-boats, one despatch-vessel, and two torpedo-boats. Of the twenty vessels already built or ordered but three are in commission. They vary so much in type that the following conventional data may perhaps be of some use (see table following), though it must be remembered that the performances stated are theoretical, except in the cases of the *Atlanta*, *Boston*, and *Dolphin*.

The defects found in the *Atlanta* when first tested were so easily remedied that the machinery finally developed a maximum horse-power which was only a little less than that required by the contract; while the *Boston* reached a maximum of 4248.5 horse-power. In his last report to the president the secretary of the navy said:

> The *Dolphin* and the *Atlanta* having both been completed, and having had trial trips, it is possible to compare them in their results with similar vessels built contemporaneously elsewhere. The *Dolphin*, of 1500 tons displacement, can be compared with the *Alacrity* and *Surprise*, English despatch-vessels of 1400 tons each, and the *Milan*, a French despatch-vessel of 1550 tons, all built contemporaneously. The *Dolphin* was designed for 2300 indicated horse-power, the *Alacrity* and *Surprise* each 3000, and the *Milan* 3900. The highest mean horse-power developed upon trial was, in the case of the *Dolphin*, less than 2200; of the *Alacrity*, 3173; of the *Surprise*, 3079; of the *Milan*, 4132. The highest speed of the *Dolphin*, resulting from several trials, was 15.11 knots, running light; of the *Alacrity*, 17.95 knots; of the *Surprise*, 17.8 knots; of the *Milan*, 18.4 knots.
>
> The *Atlanta*, the sister-ship to the *Boston*, can be compared with the *Esmeralda*, the *Giovanni Bausan*, and the *Mersey*. All three were built in England: the *Esmeralda* for Chile, the *Giovanni Bausan* for Italy, and the *Mersey* for the English government. The *Atlanta* is of 3000 tons displacement; the *Esmeralda*, 2920; the *Giovanni Bausan*, 3086; and the *Mersey*, 3550. The *Atlanta* was designed to attain an indicated horse-power of 3500, the *Esmeralda* and the *Giovanni Bausan* each 5500, and the *Mersey* 6000. The trials had of the *Atlanta* indicate that her engines will develop less than 3500 horse-power, while the *Esmeralda* developed 6000, the *Giovanni Bausan* 6680, and the *Mersey* 6626. The

maximum speed of the *Atlanta* will be less than 15 knots, while that of the *Esmeralda* was 18.28 knots, the *Giovanni Bausan* 17.5 knots, the *Mersey* 17.5 knots.

Thanks to the force of public opinion, liberal appropriations have been made for the navy. Leaving out of consideration the double-turreted Monitors, the additions to the fleet have been the cruisers *Charleston* and *Baltimore*, the No. 1 and 2 gun-boats, the cruiser *Newark*, the two armoured vessels, the torpedo-boat, the dynamite cruiser, the No. 4 and 5 cruisers, the No. 3 and 4 gun-boats of No. 1 type, and the floating batteries. The *Stiletto*, if accepted, will be bought from the Herreshoff Company; all the rest, except the battleship, are to be or have been constructed by contract in private yards. Of the new ones the *Charleston* and No. 5 cruiser will be built at San Francisco; gun-boat No. 2 at Baltimore; the dynamite cruiser, gun-boat No. 1, the *Baltimore*, *Newark*, and cruiser No. 4, at Philadelphia, and gun-boats Nos. 3 and 4 at New York.

The steel partially protected cruiser *Charleston* is, except in details of internal accommodations, a duplicate of the *Naniwa-Kan*, which was in turn a progressive development of the type-making *Esmeralda*, inasmuch as she has greater speed, more powerful armament, and superior protection to stability. The plans of the *Charleston* were bought abroad simply because they could not be made here; and notwithstanding the twopenny-ha'penny criticisms this action evoked, its wisdom has been justified. The *Charleston* has neither poop nor forecastle, and the unhampered ends give in action perfect freedom of fire for two 10-inch guns, which are mounted in low, thin-plated barbettes, situated on the ship's middle fine, at a distance of sixty feet from the bow and stern respectively. These pieces are without armour protection, except that offered against machine-gun fire by a two-inch segmental shield. Between these heavy guns a high waist stretches amidships, in which six 6-inch breech-loaders are mounted on sponsons or in projecting turrets. The secondary battery includes two 6-pounder rapid-fire guns, eight machine guns, and four above-water torpedo-tubes. The 10-inch guns must always be brought back to the fore-and-aft hue for reloading, and their ammunition is passed through steel tubes which extend below the protective deck. The engines are double-compound, situated in separate compartments, and in the *Naniwa-Kan* the type developed 7650 horse-power and 18.9 knots.

In the twin-screw cruiser *Baltimore* a longitudinal water-tight bulk-head joins the double bottom, which runs under the engine

Name or Ship.	Keel laid.	Launch.	Condition.	Material.	Displacement.	Length.	Beam.	Draught.	Speed.	Horsepower.	Main.	Secondary.	Cost.
	Date.	Date.			Tons.	Feet.	Ft. In.	Ft. In.	Knots.				Dollars.
Armored.													
Puritan	1875	1883	Completing.	Iron.	6000	280	60	18	13¼	3,500	4 of 10 inch		2,300,970
Miantonomoh	1874	"	"	"	3887	250	55.6	14.3	10	1,600	4 " 10 "		1,637,110
Amphitrite	1874	"	"	"	3887	250	55.6	14.2	10	1,600	4 " 10 "		1,530,930
Monadnock	1874	"	"	"	3887	250	55.6	14.2	10	1,600	4 " 10 "		1,592,849
Terror	1874	"	"	"	3887	250	55.6	14.2	10	1,600	4 " 10 "		1,891,077
Battle-ship			⎧ Plans under consideration.	Steel.	6000						4 " 10 " {2 " 10 "}	Not determined.	2,500,000
Cruiser			⎨ der consideration.	"	6000						2 " 10 " {2 " 12 "}	Not determined.	2,500,000
Unarmored.													
Chicago	1883	1886	Completing.	"	4500	315	48.2	20.6	15	5,000	4 " 8 " {2 " 10 "}	2 6-pdrs., 4 4·7 mm., 2 37 mm.	1,576,854
Boston	1882	1883	"	"	3000	270	42	18.6	15.5	4,248	8 " 6 " {2 " 8 "}	2 1-pdrs., 2 short Gat.	1,081,225
Atlanta	1882	1885	In commis'n.	"	3000	270	42	18.6	15.5	3,482	8 " 6 " {2 " 8 "}	2 6-pdrs., 2 3-pdrs., 2 1-pdrs.	1,081,225
Dolphin	1882	1885	"	"	1485	240	32	14.3	15	2,300	1 " 6 "	2 47 mm., 2 37 mm., 2 short Gatlings.	460,000
Charleston	1887		Building.	"	3730	300	45	19.6	18	7,500	2 " 10 " {6 " 8 "}	2 6-pdrs., 4 4·7 mm., 2 Gatlings.	1,017,500
Baltimore	1887		"	"	4413	315	48.6	21	19	10,750	4 " 8 " {6 " 6 "}	4 6-pdrs., 2 3-pdrs., 1 1-pdr.	1,325,000
Newark	1887		"	"	4083	310	49.2	20.6	18	8,500	12 " 6 "	4 37 mm., 2 short Gatlings.	1,300,000
Gun-boat No. 1	1887		"	"	1700	230	36	15	16	3,500	6 " 6 "	2 6-pdrs., 2 3-pdrs., 1 1-pdr.	455,000
Gun-boat No. 2	1887		"	"	870	175	31	12.6	13	1,300	6 " 4 "	2 37 mm., 2 sb. Gat.	247,000
Dynamite cruiser	1887		"	"		239	26.6	7.6	20	3,200	3 " 10½ " (Dynamite.)	2 3-pdrs., 1 1-pdr., 2 37 mm., 2 short Gat.	350,000
Cruiser No. 4	1887		"	"	4083	310	49.1 4/10	18.9	19	10,500	12 " 6 "	Not yet determined.	1,500,000
Cruiser No. 5	1887		"	"	4083	310	49.1 4/10	18.9	19	10,500	12 " 6 "	Not yet determined.	1,500,000
Gun-boat No. 3	1887		"	"	1700	230	36	14	16	3,500	6 " 6 "	Same as Gun-boat No. 1.	550,000
Gun-boat No. 4	1887		"	"	1700	230	36	14	16	3,500	6 " 6 "		550,000
Torpedo-boat			Not designed.		108	90	11	3	23			2 rapid-fire guns. 5 torpedoes.	100,000
Stiletto	1884	1885	Completed.	Wood.	356				22.9	560			25,000
Floating batteries			Not designed.	Steel.								Not yet determined.	2,000,000

and boilers to a protective deck that extends the whole length of the ship, and is three inches thick on the flat top and four inches thick on the sloping sides. The machinery consists of a pair of triple-expansion compound engines which are to develop 18 knots and 7500 horse-power with natural, and 19½ knots and 10,750 horse-power with forced, draft. There are two separate engine-rooms and two boiler-rooms and the normal coal capacity of 600 tons will be sufficient for 1800 knots. Additional space is provided for 300 tons more, and with this total there ought to be an endurance of 8000 miles at 11 knots, and of 14,000 miles, or 75 days' steaming, at 8 knots. No sails except storm-sails will be provided. The *Baltimore* is to have a poop and forecastle, on which four 8-inch guns with direct fore-and-aft fire will be mounted. On the main deck six 6-inch guns will be carried in broadside, and the secondary battery and torpedo-tubes are effective and well disposed.

The maximum price fixed at first by Congress for the *Newark* was less than any of the bids received, but at the last session the appropriation was increased to $1,300,000, and the contract was awarded in August of this year. The *Newark* is a bark-rigged, twin-screw cruiser of 4083 tons displacement. A double bottom extends through 129 feet of her length, and a protective deck, which rises fifteen inches above the water-line amidships, runs uninterruptedly fore and aft. Four feet above this the berth deck is built, the intermediate space being greatly subdivided and utilized for stores. Numerous water-tight frames are worked in the double bottom, and wherever practicable the cellular construction is employed. The engines are to develop 6000 horse-power with natural draft, and 8500 horse-power and a maximum speed of eighteen knots with forced draft. This vessel has a poop and forecastle, and the guns are carried on the upper deck. The main battery consists of twelve 6-inch centre-pivot guns, furnished with segmental shields, and mounted on sponsons so as to obtain the greatest arc of fire; the two guns nearest the bow and stern converge their fire at a point 400 feet distant from the ends of the ship, and those in broadside can be concentrated within 100 feet of the ship. In addition to the secondary battery-given in the table, there are six above-water torpedo-tubes.

The development of naval construction cannot be proved more conclusively than by comparing the new cruisers with those which were first laid down. In the *Atlanta*, for example, the builder guaranteed that 664 tons of machinery would produce 3500 indicated horse-power; but the *Charleston* must, before acceptance, develop 7000 horse-power for 710 tons of machinery; that is to say, the energy for weight has been doubled within four years.

The twin-screw gun-boat No. 1 is the prototype of a class that now include three vessels, and a very promising nucleus it is for a fleet to which the defence of the country's coast must mainly be entrusted. The ship is to be built of steel, with a three and a half inch complete water-tight deck, so arched as to have a spring of about three feet in its greatest width, and a crown that will nearly reach the water-line level. There is no double bottom, but the number of water-tight compartments is very great, and coffer-dams surround the engine and fire room hatches, and are carried to a height of eighteen inches above the main deck. The complement numbers 150, and the rig is that of a three-masted schooner, with a sail area of 4400 square feet. The machinery is estimated to indicate 2200 horse-power with natural draft, and 3300 with forced draft, and consists of two independent compound engines placed in separate compartments. The speed is given as sixteen knots, but it is probable this rate will be considerably exceeded. The main battery consists of six 6-inch guns, the secondary of two 57 millimetre rapid-fire guns, two 37-millimetre revolving cannons, and one short Gatling. Four of the 6-inch guns are mounted on the poop and forecastle—two forward, two aft—and the other pieces of this calibre are carried on sponsons amidships, so as to have a large arc of fire about the beam. The elevated guns are eighteen feet above the low-water line, the centre ones ten feet, and all are mounted on central pivots and fitted with protective shields. The torpedo armament is of great relative importance; of the eight tubes supplied, the stem and stern ones are fixed and fitted with under-water discharge, while the other six can be trained, and are distributed four forward and two aft. In gun-boat No. 2 the machinery is to develop 900 horse-power with natural, and 1350 with forced, draft; the engines, boilers, and magazines are placed beneath a steel deck three-eighths of an inch thick, which amidships is twenty-seven inches below the water-line at the edge and eight inches above at the crown. The armament consists of four 6-inch sponson-mounted guns, two 47-millimetre guns, two 37-millimetre revolving cannons, and one short Gatling. She is barkentine rigged, with a plain sail area of 4480 square feet, and has a slightly ram-shaped, cast-steel stem. The complement is 100.

The pneumatic-gun cruiser is to be armed with three of Zalinski's pneumatic dynamite guns of 10½-inch calibre, each of which is to throw a shell containing 200 pounds of high explosives for a distance of one mile, and to be capable of being discharged at least once in two minutes. The guaranteed speed is twenty knots.

Under the law of August 3, 1881, authorizing the construction of two new ships, it was provided that these should be:

Sea-going, double-bottomed, armoured vessels of about 6000 tons displacement, designed for a speed of at least sixteen knots an hour, with engines having all necessary appliances for working under forced draft, to have a complete torpedo outfit, and be armed in the most effective manner.

According to the circular issued by the Navy Department, one of these was to be an armoured cruiser, with a maximum draught of twenty-two feet, and the other a battleship, with a draught of twenty-three feet; both were to be built of steel, with double bottoms, to have numerous water-tight compartments fitted with powerful pumping apparatus, and to be supplied throughout with perfect drainage and ventilation. A ram bow, twin screws, electric search-lights, torpedo outfit, and a protected steel-armoured deck running the whole length of the ship and covering the boilers, engines, and magazines, were essentials; while high power and economy were so equally demanded that, to a maximum maintained speed of seventeen knots when fully equipped, great coal endurance and small fuel consumption were to be added. In each vessel a space sufficient for two hundred and seventy people, for provisions for three months, and for water for one month, was required. The cruiser was to have two-thirds sail-power on two or three masts, each supplied with a military top fitted to mount one or more machine guns. The armament of this ship was to include ten steel breech-loading rifles—four of 10-inch and six of 6-inch calibre—and a secondary battery of four 6-pounders, four 3-pounders, and two 1-pounders, rapid-fire, and four 47-millimetre and four 37-millimetre revolving cannons, all of the Hotchkiss pattern, together with four Gatling guns. There were to be fitted six torpedo-tubes—one bow, one stern, and two on each side, of which at least one on each side forward was to be under water. The heavy guns were to load in not less than two positions, and were to be protected by at least ten and a half inches of steel armour, properly backed; the 6-inch guns were to be fitted with shields, and all the guns were to be arranged so as to obtain the greatest horizontal and vertical fire consistent with other conditions. Any vertical armoured protection at the water-line was to be at least eleven inches thick in the heaviest part, and thicker, if practicable.

The armament of the line-of-battle ship was to consist of two 12-inch and six 6-inch guns, and of a secondary battery which included four 6-inch, six 3-pounder, and two 1-pounder rapid-fire guns; of four 47-millimetre and four 37-millimetre revolving cannons, and of four Gatlings. The torpedo outfit was similar to that of the cruiser.

The plans submitted were opened on April 1st of this year, and notwithstanding the difficulties which the displacement imposed upon the other requirements, no less than thirteen designs were received from ten different competitors. The most important of these were offered by the Thames Iron Ship Building Company and the Barrow Ship Building Company, of Great Britain; by A. H. Grandjean, Esq , of France; and by chief constructor Wilson, naval constructor Pook, and Lieutenant Chambers, all of the United States navy. The designs were submitted to a board, and this finally recommended the Barrow plan as best suited for the armoured battleship. So far as the armoured cruiser was concerned, the Board reported as follows:

> The marked differences in the essential features of the designs of armoured cruisers of the Barrow Ship Building Company, Lieutenant W. I. Chambers, A. H. Grandjean, and the Thames Iron Works and Ship Building Company, prevent their classification in the order of merit. Each exhibits features which strongly commend themselves, but the Board does not consider it advisable for the government to build a vessel upon any one of these plans.

The battleship, though designed by one of the most distinguished marine architects in England, has not in its present form received the general approval of experts, for between it and the plan submitted by the Bureau of Construction there seem to be differences of merits which are strongly in favour of the latter. The dimensions of the new ships are as follows:

BARROW SHIP	NAVY DEPARTMENT SHIP
Length between perpendiculars, 290 feet; on load water-line, 300 feet; extreme breadth, 64 feet 1 inch; mean draught, 22 feet 6 inches; displacement, 6800 tons.	Length between perpendiculars, 300 feet; on load water-line, 310 feet; extreme breadth, 58 feet; mean draught, 22 feet; displacement, 6600 tons.

The striking differences between these two ships are found in their relative stability and sea-going qualities. Mr. John, the designer of the Barrow ship, in a paper on Atlantic steamers, read before the Institution of Naval Architects July 29, 1886, made the following statements:

> This question of stability will have to be carefully watched and studied within the next few years, because there is a tendency at present towards a rapid increase in the proportion of beam to length; and as the draught of water in these large ships is limited,

we must be careful that in seeking higher speeds with increased beam we do not get too much stability, and so render the vessels heavy rollers and very uncomfortable as passenger-ships. It is possible the future may see vessels of greater beam than any yet afloat in the merchant-service; but if so, it is almost inevitable that they will have to be made higher out of water in order to render them easy and comfortable at sea, but even that has its limits. Perhaps it is well to give an extreme case, and here I will make use of our old friend *The Great Eastern*.... Now, for the purpose of trading it is quite clear that *The Great Eastern* cannot be loaded much deeper than other ships, while her beam is half as great again; and the consequence is, her stability, as compared with our modern passenger-ships, is so excessive that she is bound to be a tremendous roller among the heavy seas in the Atlantic. Her metacentric height, when loaded, was, I believe, stated by the late Mr. Froude to be as much as 8.7 feet, which is from three to four times as much as is thought sufficient for ships in the present day, or consistent with their easy behaviour at sea.

Thus Mr. John himself regards 2.9 feet to 2.2 feet as the proper metacentric height for those steamers, and it is generally considered by modern designers that from 2.5 to 3.2 feet is most suitable for this class of armoured ships, and is conducive to easiness of motion in a sea-way. The value of this quality to a ship intended for sea-fighting cannot be overestimated, for upon her steadiness as a gun-platform the aim and efficiency of her guns greatly depend.

It will be noticed that this ship has exceptionally great beam, that of most ships of her class and displacement, varying from 54 to 59 feet, and judging from the sketches which have appeared, her water-line co-efficient is about 0.72. From an approximate calculation based on this assumption it is found that her metacentric height will be about six feet. The water-line coefficient may possibly be a little finer than 0.72, and thus reduce the metacentric height, but if this ship is assumed to have a metacentric height of three feet, her water-line coefficient would be 0.6288, which is an *impossibility*, if her coefficient of fineness of displacement be that given in the published dimensions. Such a water-line and coefficient of fineness for 6300 tons displacement would produce a perfect rectangle for a midship section. So that, unless her dimensions are changed, she will surely be a heavy roller, and after much sea duty she will suffer such severe strains as to require frequent and costly repairs.

The battleship designed at the Navy Department has very different qualities, if the dimensions already published be correct. To possess a metacentric height of three feet she would require a water-line coefficient of 0.753, and a midship-section coefficient of 0.89 to 0.90, which is a good proportion for such a vessel. Not only in sea-going qualities does the American design seem to be superior, but her battery is far more powerful and better disposed in every way, while her speed and endurance are equally as great as the plan recommended. Mr. John has adopted the *echelon* arrangement of heavy guns, a disposition which both the English and Italian governments have, after long trial, discarded in their latest ships. When the first sketches of a design are made, this arrangement of guns is theoretically perfect, as it is supposed to give quite as much power of fire ahead and astern as on each broad-side; but when the design is developed and practically tested, it is found that too much of the ship's efficiency in other respects is sacrificed, that the powerful end fire is not attained, and that the broadside is greatly weakened, owing to the obstructed arcs of fire.

Besides this, the guns, being placed at some distance from the midship line, have less accurate fire in rolling, and the ship's propensities to roll are encouraged and are greater than would be the case if the guns were placed on the midship line. It is also found that the blast from the heavy guns is destructive to superstructures and other fittings on the upper deck. The Italians, indeed, have placed stout ventilating shafts on their *Italia* and *Lepanto* to prevent the rearmost pair of heavy guns from being trained within twenty degrees of the fore and aft line. This is done so that the blast from these guns will not prostrate the gunners attending the other pair, notwithstanding the fact that those men are under the armour cover. The *Duilio*'s forward smoke-pipe is placed entirely on the port side of the fore and aft line, in order to permit of one pair of turret guns firing ahead. The upper-deck, 6-inch, central-pivot guns of the *Andrea Doria* class are now to be placed wholly within the superstructure, in order to be out of danger from the blast of the heavy guns when the latter are fired near the line of keel, and the same change would have to be made with the upper-deck, 6-inch guns in the Barrow design.

Similar objections exist to the Bureau of Construction design for an armoured cruiser. This vessel, although possessing the bad features inherent in the *echelon* arrangement of heavy guns, does not have the best ideas of the Barrow design, *i.e.*, high freeboard, heavy guns mounted high above the water-line, and commodious quarters for officers and men. Both designs besides have the very objectionable and old-fashioned features of requiring the turrets to be revolved to fixed loading positions after being fired. The Bureau cruiser, it may be said, is not saddled with too much metacentric height. She has ten feet less beam, her centre of gravity is about one foot lower, and unless her water-line coefficient is very full, she will have a metacentric height rather less than what is regarded to be the best.

It is not surprising, however, that the Bureau plans are so different in efficiency, for while the better plan, the battleship, is original with the Navy Department, the armoured cruiser is a copy of, and no substantial improvement over, that of the Brazilian ship *Riachuelo* designed several years ago. This ship is considered one of the best of her date, but great improvements in ship design have been made within the past few years, and it is against the tendencies of American inventive genius to take a step backward.

The general plans of cruisers No. 4 and 5 were published in the *New York Herald* of June 1st, together with the following data:

> They are to be twin-screw cruisers, 310 feet long on the water-line, 49 feet 1¾ inches extreme breadth, 18 feet 9 inches mean draught, displacing 4083 tons. They are to have machinery of 10,500 indicated horse-power under forced draft. The maximum speed is 19 knots, rig that of a three-masted schooner, spreading 5400 square feet of sail. They will have a double bottom extending through 129 feet of the length. The framing in this portion is on the bracket system. Before and abaft the double bottom, above the protective deck, Z-bars form the transverse frames. The protective deck, which is nineteen inches above the water-line amidships, is flat across the top, with sides which slope down to a depth of four feet three inches below the water-line. The horizontal portion is two inches thick, the slope being three inches, reduced at both ends to one and a half inches. It extends uninterruptedly forward and aft, and protects the machinery, magazines, and steering-gear, the machinery being further defended by the disposition of the coal-bunkers. The main hatches in this deck are protected by armour-bars, and have coffer-dams extending to the upper deck. The guns are carried on the gun, forecastle, and poop decks.
>
> *Armament.*—The main battery, which consists of twelve 6-inch breech-loading rifles, all on centre-pivot mounts, with two-inch segmental steel shields, is arranged on sponsons so as to obtain the greatest possible arc of fire. The forecastle, the

poop, and the bridges have been as much as possible availed of to shelter the guns. The two guns for-ward and the two guns aft converge their fire a short distance from the ends of the ship, and the broadside can be concentrated within 100 feet of the side. Four above-water torpedo-tubes are provided on the berth-deck, and two direct ahead under-water torpedoes in the bow. The secondary battery is composed of four 4:7-millimetre revolvers, four 57-millimetre single-shots, two 37-millimetre revolvers, and one short Gatling. The coal capacity is 850 tons. The complement of men 300. . . .

To appreciate what is required to make nineteen knots an hour at sea, we have only to remember that the *Umbria* and *Etruria* are 500 feet long, with more than 12,000 tons displacement and 14,500 indicated horse-power, ordinarily making 18½ and on special occasions 19 knots an hour. Now, to increase her speed to 20 knots an hour, the *Umbria* would require about 19,500 horse-power, which means 5000 extra horse-power for the extra knot. For a second extra knot would be required about 6000 horse-power more, making about 25,000 horse-powder necessary to develop a speed of 21 knots.

Gun-boats Nos. 3 and 4 are to be copies of gun-boat No. 1. No designs for the floating batteries and the torpedo-boat have been published. The *Stiletto* is one of the famous Herreshoff boats, and is now being tested in consequence of a favourable report made by a board of officers. On July 23, 1886, with a total displacement of twenty-eight tons, she made an average of 22.12 knots as the mean of four runs over the measured mile in a rough sea and fresh wind, and on July 30th she attained an average of 22.89 knots. These were excellent results for a boat ninety feet in length, and promised that the type, with certain modifications, was equal to greater demands. The trial data of this year have not yet been published, though it is unofficially reported that her performance was equally as creditable.

United States Naval Artillery

From the time of the introduction of cast-iron cannons in 1558 until a comparatively late period, development in naval artillery proceeded at a very slow rate. The security that was attained by the adoption of cast-iron was so great, as compared with the danger attending the use of the more ancient artillery, that the new guns were regarded as fully supplying all the demands of a suitable battery. The guns were muzzle-loaders, making the manipulation simple, the previous rude attempts at breech-loading being abandoned. The number of calibres

BRONZE BREECH-LOADING CANNON CAPTURED IN COREA

that were introduced was very numerous, partly to suit the weight of the batteries to the ships, and partly to accommodate the fancy of the time for placing in different parts of the ships guns varying much in size and destructive effect. The general character of the batteries and the multiplication of calibres can best be illustrated by noting the armament of two typical ships of the seventeenth century.

The *Royal Prince*, a British ship built in 1610, carried fifty-five guns. Of these, two were *cannon-petronel*, or 24-pounders; six were *demi*-cannon, medium 32-pounders; twelve were *culverins*, 18-pounders, which were nine feet long; eighteen were *demi-culverins*, nine-pounders; thirteen were *rakers*, 5-pounders, six feet long; and four were *port-pieces*, probably swivels. These guns were disposed as follows: on the lower gun-deck, two 24-pounders, six medium 32-pounders, and twelve 18-pounders; on the upper gun-deck the battery was entirely of 9-pounders; and the forecastle and quarter-deck were armed with 5-pounders, and the brood of smaller pieces which swelled the nominal armament.

The *Sovereign of the Seas*, built in 1637, in the reign of Charles I., was unequalled by any ship afloat in her time. She mounted on three gun-decks eighty-six guns. On the lower deck were thirty long 24-pounders and medium 32-pounders; on her middle deck, thirty 12-pounders and 9-pounders; on the upper gun-deck, "other lighter ordnance;" and on her quarter-deck and forecastle, "numbers of murdering pieces."

In the obstinately contested actions between Blake and Van Tromp in the Cromwellian time, the ships and batteries did not differ in any great degree from those contemporaneous in construction with the *Sovereign of the Seas;* and when we remember the inferior character of the powder used in those days we can account for the duration of some of the engagements between the English and Dutch ships which were sometimes protracted through three days.

The brood of "murdering pieces" of small calibre and little energy was, after many years, dispersed by the introduction of carronades—a short cannon of large calibre, which was found to be a convenient substitute for the 8-pounders and 9-pounders on upper decks, and for the "lighter ordnance," which was ineffective; but this change was brought about slowly, as is seen by referring to the batteries of some ships which fought at Trafalgar.

The Spanish seventy-fours in that action had fifty-eight long 24-pounders on the gun-decks; on the spar-deck, ten iron 36-pounder carronades and four long 8-pounders; and on the poop, six iron 24-pounder carronades—total, seventy-eight guns.

Bronze breech loaded used by Cortez in Mexico

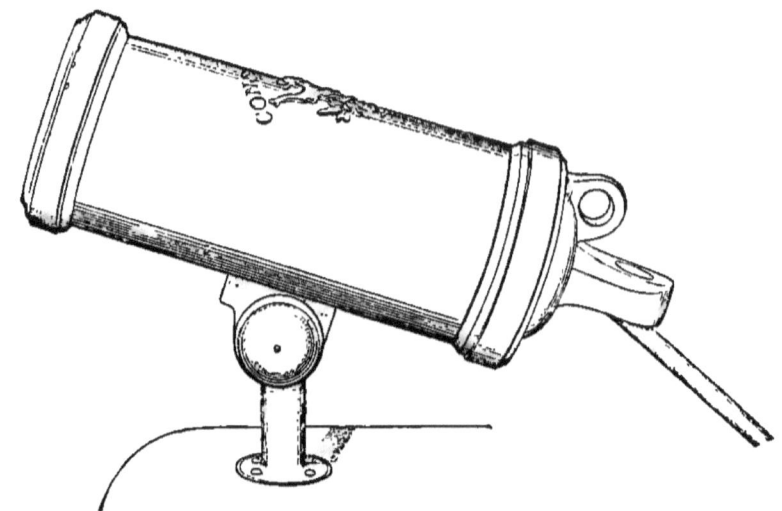

Breech-loader captured in the war with Mexico

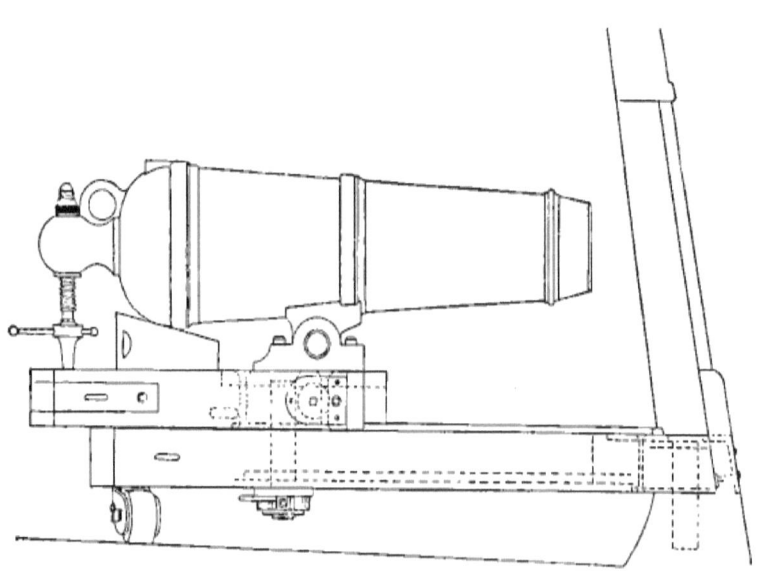

USN carronade, slide & carriage

The *Victory*, the English flag-ship, mounted on her three gun-decks ninety long 32, 24, and 12 pounders, and on the quarter-deck and forecastle, ten long 12-pounders and two 68-pounder carronades.

The *Santissima Trinidada* mounted on the lower gun-deck thirty long 36-pounders; on the second deck, thirty-two long 18-pounders; on the third deck thirty-two long 12-pounders; and on the spar-deck, thirty-two 8-pounders. In the British accounts she is said to have had one hundred and forty guns, which number must have included swivels mounted for the occasion.

At the end of the eighteenth century the 18-pounder was the preferred gun for the main-deck batteries of frigates, guns of larger calibre being found only on the lower decks of line-of-battle ships. The 18-pounder was the maximum calibre that was employed on board the ships of the United Colonies of North America in the war of the Revolution. The resources of the colonies did not admit of building ships to contend with vessels fit to take their place in line of battle, but such as were constructed were well adapted to resist the small British cruisers, and to capture transports and store-ships. The so-called frigates of that day were vessels varying from six hundred to a thousand tons, and, according to their capacity, carried 12-pounders or 18-pounders in the main-deck batteries. There was usually no spar-deck, but the forecastle and quarter-deck, which were connected by gangways with gratings over the intermediate space, were provided with an armament of light 6, 9, or 12 pounders. A few carronades came into use during this war.

At the conclusion of this war the Colonial fleet disappeared, and it was not until the time of the depredations on the growing commerce of the United States by the Algerine corsairs that Congress felt justified in incurring the expense of establishing a national marine. The ships which were built under the law of 1794 were fully up to the most advanced ideas of the time, and some of these ships carried on their gun-decks a full battery of 24-pounders, thirty in number, while the others were armed with 18-pounders on the gun-deck, with spar-deck batteries of 9 and 12 pounders, the carronade not having been yet definitely adopted for spar-deck batteries.

It is not until the war of 1812 that we find the carronade fully established as the spar-deck armament of frigates. The *Constitution* and the *Guerrière* carried 32-pounder carronades of very similar weight and power in the place of the long guns of smaller calibre on the spar-deck. The original name of this piece of ordnance was the "Smasher," the leading purpose of the inventor. General Melville, of the British

Bronze 12-pounder, El Neptuno, 1781

artillery, being to fire 68-pounder shot with a low charge, thus effecting a greater destruction in a ship's timbers by the increased splintering which this practice was known to produce. Carronades of small calibre were subsequently cast, which were adopted for spar-deck batteries of frigates and line-of-battle ships, and, as they grew in favour, formed the entire battery of sloops-of-war and smaller vessels until about 1840, when the attention that had been given for some years to the subject of naval ordnance began to assume tangible shape, and the effort was made to proceed in this matter in accordance with an intelligent system.

The advantage of large calibre was firmly impressed upon those who occupied themselves with the ordnance matters of the navy. As the fleet was developed, the 24-pounder gave way to the 32-pounder, and for the lower-deck battery of line-of-battle ships the 42-pounder was introduced. Some 42-pounder carronades were also introduced as spar-deck batteries for these larger ships. With the disappearance of this class of ship the 42-pounder was abandoned, and the 32-pounder was retained as the maximum calibre, different classes being assigned to different sizes of ships. These classes were divided into the gun proper, with 150 pounds of metal to one of shot; the double-fortified gun, with 200 pounds of metal to one of shot; and the medium gun, with 100 pounds of metal to one of shot. The carronade of the same calibre, mounted on a slide, had a proportional weight of 65 pounds of metal to one of shot.

In the interval between 1840 and 1845 the double-fortified 32-pounder was replaced by a gun of the same calibre of 57 hundred-weight, called the long 32-pounder; and to suit the capacity of the different classes of ships then in the service, there were introduced the 32-pounders of 46 hundred-weight, 42 hundred-weight, and 27 hundred-weight, in addition to the regular medium gun of 32 hundred-weight. This period also marks the introduction of shell guns as part of the battery.

To this time no explosive projectiles had been used with cannons properly so called; their use had been limited to mortars and howitzers. The mortar was originally used for projecting huge balls of stone at high angles. The first practical use made of them for projecting bombs was in 1624, but the unwieldy weight of the mortar and its bomb, the latter sometimes exceeding 300 pounds, prevented their use in field operations. To provide for this, light mortars were cast, which, being mounted on wheels, were denominated howitzers. Frederick the Great of Prussia brought this form of artillery to its highest develop-

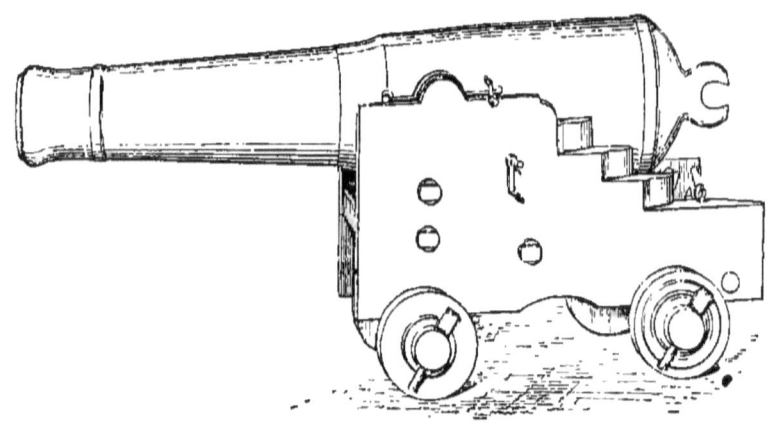

USN Medium 32-pounder

ment for field and siege use, and the Continental powers of Europe adopted it to a large extent for projecting bombs at high angles of fire. The mortar has never had a place in regular naval armaments; it has been used afloat for bombardment of cities and fortified positions, but never with a view to contending with ships.

The success attending the use of explosive projectiles at high elevations did not lead at once to their application to horizontal firing from cannons. An important link in the progress of the idea resulted from the effort to avail of the advantage of ricochet firing with bombs. In order to effect this, the angle of elevation had to be reduced to enable the bomb to roll along the ground. The reduced angle of elevation was still greater than that used for cannon, but the success of the experiment led to the casting by the French of an 8-inch siege howitzer, which, in connection with the development in the manufacture of fuses, made it practicable to apply the idea of firing shells, like shot, horizontally, and the chief object in view seems to have been to operate against ships.

The combining of the elements necessary for the achievement of this important step in naval artillery is by common consent credited to General Paixhan, of the French artillery, who, though not claiming the invention of any of the numerous details involved in the system, succeeded in so judiciously arranging the parts as to make the system practicable by which the whole character of naval armaments was revolutionized.

Following the progressive ideas of the age, shell-guns were introduced in the United States navy. These were of 8-inch calibre, and of weights of 63 hundred-weight and 55 hundred-weight. The guns were shaped in accordance with the form adopted by General

Paixhan, and were easily distinguishable in the battery from the ordinary shot-gun. From this circumstance they obtained the title of Paixhan-guns, though there was nothing special in the gun itself to merit an appellation. The whole system was Paixhan's; the gun was only a part of the system.

It required many years to bring the shell-gun into such general application as to displace the solid-shot gun. They were assigned tentatively to ships in commission, and in 1853, by a navy regulation, the battery of a frigate was provided with only ten of these guns, which were collected in one division on the gun-deck. The first vessel in the United States navy whose battery was composed exclusively of shell-guns was the sloop-of-war *Portsmouth*, in 1856. This vessel carried a battery of sixteen 8-inch shell-guns of 63 hundred-weight. These were among the first of a new pattern of gun for which the navy is indebted to the skill and study of the late Rear-admiral Dahlgren.

The determination of the best form for cannons was a question which had occupied the minds of artillerists for some years. In the older guns the thickness of metal was badly distributed; it was too uniformly extended along the entire length, not arranged in such proportions as to accord with the differing strains along the bore. Colonel Bumford, of the United States Ordnance, had been among the first to consider this subject, and for many years the results of his experiments had guided construction to a great degree. General Paixhan made a further step in advance by reducing very much the thickness of metal along the chase of his guns, but it remained for Rear-admiral Dahlgren to produce the perfection of form in the gun so widely known bearing his name. In this gun the thickness of metal is proportioned to the effort of the gases in the bore, and all projections and angular changes of form are suppressed, giving to all parts a curved and rounded surface. The suppression of angular formations on the exterior of a casting has a remarkable effect on the arrangement of the crystals while cooling. These arrange themselves normal to the cooling waves, which, if entering from directions not radial with the cylindrical casting, produce confusion in their arrangement, establishing planes of weakness where the waves meet, which, in case of overstrain on the piece, assist rupture and determine the course of the fracture.

With the introduction of the Dahlgren shell-gun the transition of the artillery of the United States navy may be said to have been completed. The shell-gun of 9-inch and 11-inch calibres followed the 8-inch, and ships were armed with such as were appropriate to their capacity as rapidly as the new guns could be manufactured. When

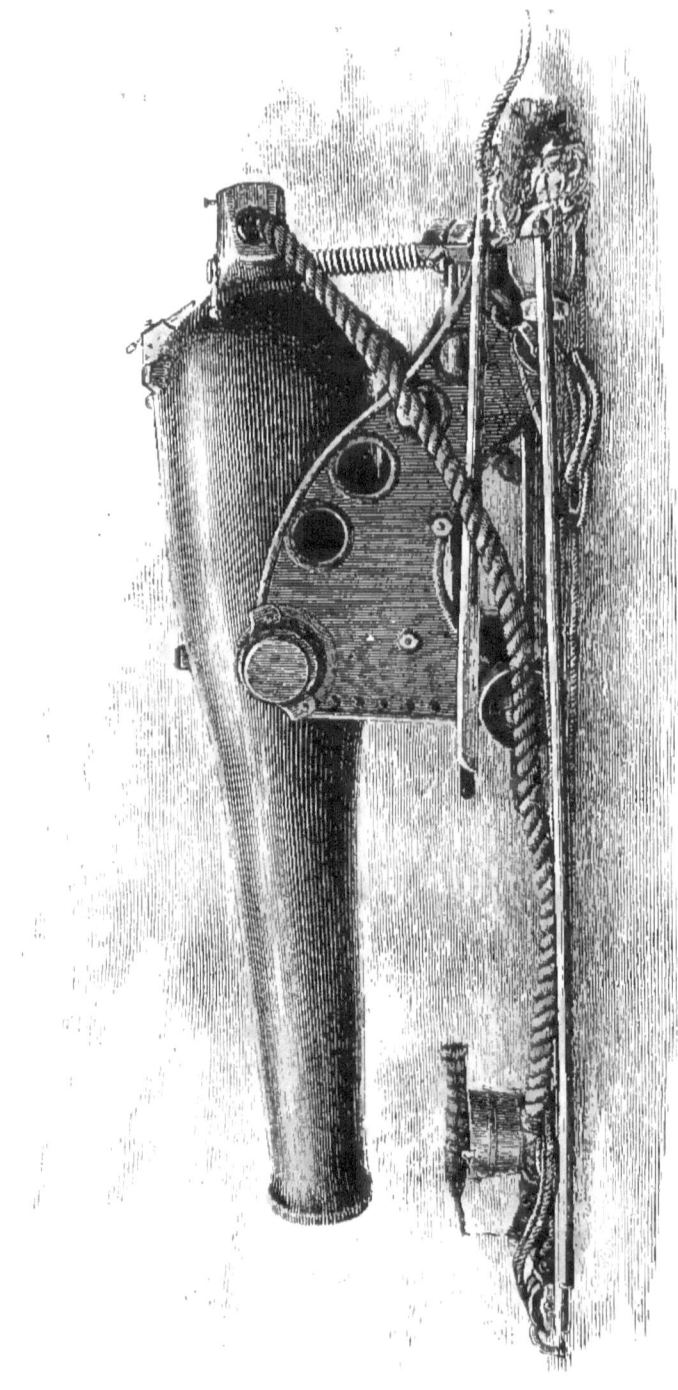

USN. 9-Inch Dahlgren (9-inch smooth-bore)

fully equipped, the armament of the United States navy was superior to that of any other navy in the world.

The substitution of shells for solid shot marks an important epoch in naval artillery. The probable effect of a shot could be predetermined and provided for; that of a shell was unknown. In order to produce serious injury with a shot, it was necessary to perforate the side of an enemy. This was not indispensable with a shell; with the latter, perforation might be dispensed with, as penetration to such a depth as would give efficacy to the explosion might prove more destructive to the hull than would absolute perforation. With the shot, damage was done to life and material in detail; with the shell, if successfully applied, destruction was threatened to the entire fabric, with all it contained. Naval artillery entered a new phase; the rough appliances of the past would no longer answer all demands. The founder could not alone equip the battery; the laboratory was called into use, and pressed to provide from its devices. The "new arm" depended upon the successful working of the fuse of the shell, without which it was but a hollow substitute for a solid shot, and this detail demanded the utmost care in preparation. It was the perfecting of this device which, more than aught else, delayed the general adoption of the new artillery for so long a time after its advantages had been recognized.

The fuses that were used to explode the ancient bombs were long wooden plugs, bored cylindrically, and filled with powder condensed by tamping it to a hard consistency. The fuse case projected from the bomb, and to avoid being bent by the shock of discharge, was placed carefully in the axis of fire. Before the discharge of the mortar the fuse was lighted by a match. In applying the fuse to shell-guns fired horizontally, the problem was so to arrange it as to ignite it by the flame of discharge, and so to support it in the wall of the shell as to prevent any dislocation of the fuse composition, the cracking of which would permit the penetration of the flame into the mass. This was successfully accomplished, and the United States navy fuse was justly famous, one feature of it being a simple but most effective device called a "water-cap," which guarded against the injurious introduction of sand or water when the shell was fired *en ricochet*. The introduction of a safety-plug in the bottom of the fuse case, which required the shock of discharge to displace it in order to open a way of communication between the fuse and the bursting charge in the shell, and the absence of all accidents in manipulation, inspired such confidence that the new arm advanced to favour, and both officers and men were proud to be identified with it.

Previous to the introduction of shells there had been in use incendiary projectiles, not explosive, but intended to set fire to an enemy's vessel. Hot shot were applied to this purpose, but the use of these was chiefly confined to shore batteries, where a suitable heating furnace could be conveniently provided. The projectile for this purpose chiefly used from ships was the carcass, which was a shot in which several radial cylindrical holes were formed which were filled with powder tamped to a hard consistency; these columns of composition were ignited by the flame of discharge, and continued to burn until consumed. The flame issuing from these holes served to ignite consumable material in their vicinity. The chief danger from a carcass was from lodgement in the side of a ship; if it landed on deck it could be removed and thrown overboard, as there was no danger from explosion; the addition of the bursting charge in the cavity of a shell produced a projectile which was far in advance both for generating a flame and for preventing interference with its mission.

The probable destructive effect of shells exploding in the sides or on the open decks of ships was thoroughly recognized, and experiments at targets sufficiently proved it; but circumstances on a proving-ground and in action are so dissimilar that the experience of a naval engagement was looked forward to with much interest, in order to satisfy as to the effect of the new projectile in all the varying conditions of a sea-fight. Referring to the history of the past thirty years, which marks the period of the general introduction of shell-guns, it is remark-able how few engagements between ships have taken place; but on every occasion of the use of shells, when unarmoured vessels were engaged, the effect has been most decided and complete. Three instances only can be referred to of purely sea-fights, *viz.*, the engagement between the Russian and Turkish fleets at Sinope in 1853, during the Crimean war, the engagement between the United States steamer *Hatteras* and the Confederate cruiser *Alabama* during the war of the rebellion, and the fight between the *Kearsarge* and the *Alabama* during the same war. In the affair at Sinope the Russian ships used shells; the Turkish had only solid shot. The result was the total destruction of the Turkish force. Not one ship escaped; all were burned or sunk. The fight between the *Alabama* and the *Hatteras* resulted in the sinking of the *Hatteras*; and the contest between the *Alabama* and the *Kearsarge* ended the career of the *Alabama*. And it may be noticed that but for the failure to explode of a shell that was embedded in the stern-post of the *Kearsarge*, that vessel might have accompanied her antagonist to the bottom of the sea.

The gallant attempt of Rear-admiral Lyons with the British wooden fleet before the forts of Sebastopol is an instance which proved the uselessness of subjecting unarmoured vessels to the steady fire of fortified positions using shells from their batteries.

One other instance of a sea-fight can be cited in the engagement in 1879 between two Chilean armoured vessels and the lightly armoured Peruvian turreted vessel *Huascar*. The *Huascar* was terribly over-matched during this fight, but at its conclusion her boilers and engines were intact, and indentations on her sides showed that her light armour had deflected a number of projectiles; but the effect of the shells that had burst on board of her was apparent in the great destruction of life.

The very decisive engagement which took place at Lissa in 1866, between the Austrian and Italian fleets, should not be omitted in alluding to sea-fights of a late period; but this action can hardly be quoted as one in which the element of shell-fire can be recognized as the exclusive cause of destruction, for the remarkable impetuosity and dash of the attack and the desperate use of the ram produced a crisis which obviated the necessity for continuous bombardment with cannon.

The necessity of providing a defence against shells was recognized both by England and France during the Crimean war, and a protection of armour was supplied to some floating batteries built at that time which were intended to operate before fortified positions; and at the conclusion of the war the English built the *Warrior* and the French built *La Gloire*. These were the first specimens of ironclad ships of war. They were capable of resisting successfully the entrance of shells from guns of the period. It is thus seen that almost coincident with the general adoption of horizontal shell-firing, naval construction entered a new phase, and a new problem was submitted to the naval artillerist.

Against an iron-faced target the solid shot might be partially effective, but the impact of the spherical shell was harmless, and the explosive effect of the bursting charge enclosed in it would be superficial. This was amply demonstrated in actual practice during our war experience, notably at Mobile Bar, in the engagement with the Confederate ironclad *Tennessee*, the roughly constructed armour of which vessel resisted a storm of our heaviest shells.

The impotency of the spherical shell against armour being recognized by foreign governments, they proceeded to develop the rifled cannon, which with its elongated projectile offered the means of effecting the object of the time—to perforate armour with an explosive projectile. Our authorities, however, persevered in their faith

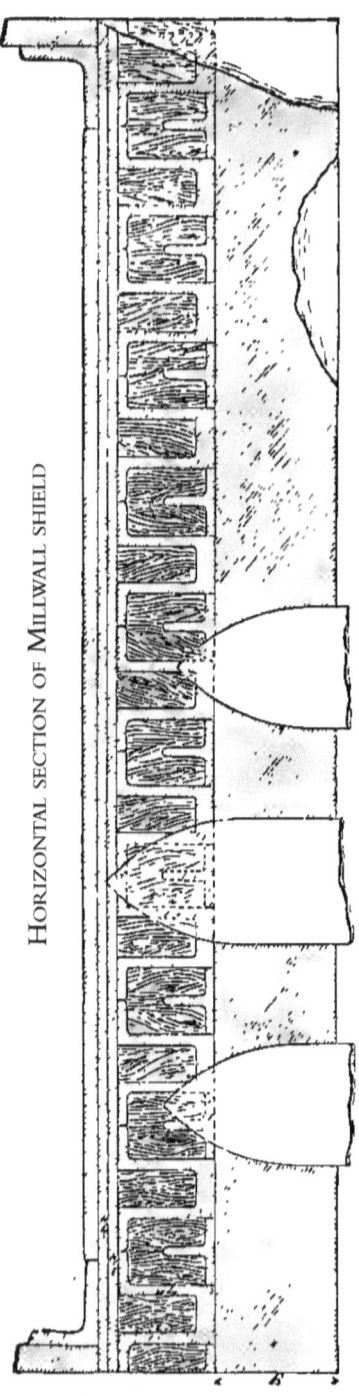

HORIZONTAL SECTION OF MILLWALL SHIELD

in the smooth-bore, and held that the *racking* effect of a spherical projectile of sufficiently large calibre was superior to that produced by the perforation of a rifle projectile of inferior diameter. The 15-inch and 20-inch smooth-bore cannons were cast in accordance with this idea, and the racking side of the question was so obstinately held that the British government imported in 1867 from the United States a 15-inch gun for the purpose of determining by their own experiments what foundation there was for the advantages that were claimed for it.

The gun was bought of Charles Alger & Co., of Boston; it weighed nineteen tons, and threw a cast-iron spherical solid shot of about four hundred and fifty pounds. It was mounted at Shoeburyness, and was fired in competition with English rifled cannons of 9-inch and 10-inch calibres. The result of the experiments went to show that against a target with a power of resistance inferior to the energy of the projectile the effect of the large sphere at short range is more disastrous than that of the elongated rifle projectile of the same weight; but that against a target able to resist the total energy of both the injury done by the rifle projectile is by far the greater. The comparative effect is well shown on a target called the "Millwall Shield," consisting of a plate nine inches in thickness, backed by Hughes's hollow stringers—an arrangement of target which to the time of the experiment had proved invincible. The 15-inch smooth-bore spherical shot rebounded from the target six feet, leaving a 3-inch indentation on the plate, while the 9-inch rifle projectile, weighing two hundred and fifty pounds, made complete penetration of the plate, passing two or three inches into the backing, and the 10-inch rifle projectile, weighing four hundred pounds, penetrated to the rear of the backing itself.

It should be mentioned in this connection that the United States government adopted during the war of the rebellion a rifled cannon proposed by Captain Parrott of the West Point Foundry, New York, of which many were introduced into both the navy and army, and did good service as long as the charges of powder were limited in weight; but when these guns were called upon for work requiring great endurance, they proved untrustworthy and dangerous to those who served them. At the naval bombardment of Fort Fisher several of them burst, causing loss of life on board the vessels of which they formed the armament. They were constructed of cast-iron, having a coiled hoop of wrought-iron shrunk around the breech. They have ceased to form a part of our naval armament.

During the years of inaction in the United States that have intervened since these experiments, the smooth-bore partisans have had time to reflect and to learn lessons of practical usefulness from observing what has been transpiring abroad. Opportunities have been afforded to note the progress made in armour and artillery, and though the smooth-bore shell is still operative against unarmoured vessels, the advantages of the rifled gun under all the circumstances of navy experiences have been admitted, and in the transition through which our naval artillery is now passing we are not embarrassed by the presentation of views antagonistic to the principles on which it has been determined our new artillery is to be constructed. The system at the basis of our present acts is founded on a comprehensive view of the whole subject, and is intended to provide our ships with a surplus of offensive power over what their capacity for defence might seem to call for.

Our navy will possess a certain number of armoured vessels for coast defence, and armoured sea-cruisers are certain to be included in the list, but the more numerous class will be unarmoured, and the first problem to be solved is that of providing for these a suitable armament.

The work to be done by an unarmoured cruiser must be done from a distance when risking an engagement with an armoured en-

A Krupp gun on a naval carriage

emy. The superiority of armament must compensate for deficiency in defensive power which precludes close quarters. To make these ships effective they must be armed with guns capable of doing an extraordinary amount of work, and yet the size of the vessels will not admit of their carrying guns of immense weight. In order to get this amount of work out of a comparatively light gun, we must secure great initial velocity for the projectile. This can only be done by burning a large charge of powder, which involves a long bore in which to burn it, while care is necessary to secure a large margin of strength in the material of which the gun is constructed. These essential demands required a radical change in the form and material of our present armament; they also forced a change in the method of construction.

The superior fitness for cannons of steel over cast-iron was recognized many years ago, but the difficulty of casting steel in large masses prevented the introduction of steel guns, and the generally acceptable treatment of cast-iron made it answer satisfactorily the demands for gun-metal not subjected to unusual strains. Frederick Krupp, of Essen, in Germany, was the first steel manufacturer who succeeded in casting steel in large masses, and he produced a number of steel guns cast from crucibles in solid ingots, which were bored, turned, and fashioned as in the case of cast-iron smooth-bore guns. These guns held a position in advance of other manufactures on the score of strength of material. But the introduction of the rifle system, the call for higher velocities, the increased charges of powder, with the consequent increase of strain, enhanced by the friction attending the passage of the projectile forced along the bore, had the effect of calling attention to the weakness that was inherent in the method of construction of cannons.

It is well known that an explosive force operating in the interior of a hollow cylinder of any thickness is not felt equally throughout the wall of metal; the parts near the seat of explosion are called upon to do much more work in restraining the force generated than are the parts more remote. It has been determined that the strain brought upon the portions of the wall is in inverse proportion to the squares of their distances from the seat of effort. Thus, in a gun cast solid, if we take a point two inches from the bore, and another four inches from the bore, the strain felt at those points respectively will be inversely in the proportion of four to sixteen, or, in other words, the metal at two inches from the bore will be strained four times as much as that at the distance of four inches. From this it can be seen that the metal near the seat of effort may be strained beyond its tensile strength, while that more distant is only in partial sympathy with it. Rupture thus

Alfred Krupp

originates at the interior portion, and the rest of the wall yields in detail. No additional strength of material can change this relationship between the parts; they result from a law, and show that this method of construction for a cannon is untrustworthy where the strains approach the tensile strength of the material.

The means of providing against this successive rupture of overstrained parts is found in the "built-up gun," in which an interior tube is surrounded by encircling hoops of metal, which are shrunk on at sufficient tension to compress the portions which they enclose. This is the principle of "initial tension," which is the basis of the modern construction of cannons. By adopting this method, an ingot to form a tube to burn the required amount of powder can be cast of a light weight in comparison with what would be needed for a complete gun, and the strength and number of reinforcing rings to be shrunk around it can be readily determined, proportioned to the known strain that will be brought upon the bore of the piece. The late developments in the manufacture of steel by the open-hearth process remove all difficulty to procuring the necessary metal in masses suitable for all parts of the heaviest guns.

The built-up steel gun is the one now adopted in Europe by the leading powers, and it is the gun with which the United States navy will be armed; but, before its final adoption, efforts were made to convert old smooth-bore cast-iron guns into rifles, and to construct new guns partially of steel and partly of wrought-iron. As some of these methods of conversion offered an economical means of acquiring rifled cannons, our naval authorities were led into the error of countenancing the effort to a moderate degree.

The system that was adopted was that originally suggested by Mr. P. M. Parsons in England, which was afterwards patented by Major Palliser, R. A., and bears his name. It consisted in enlarging the bore of a cast-iron gun, and inserting a tube of wrought-iron formed of a bar arranged in the form of a coil when heated. The tube was expanded by firing charges of powder, and afterwards rifled. The guns are muzzle-loaders, and are not increased in length beyond that of the cast-iron gun which forms the casing for the tube. The length is thus limited in order to preserve the preponderance of the piece, and because of the want of longitudinal strength in the coil, which cannot be depended on beyond a few tons' strain; the arrangement of metal in a coil provides very well for circumferential or tangential strains, but in the Palliser conversion the longitudinal strength depends on the cast-iron casing. The idea of the coiled wrought-iron tube originated

BREECH-LOADING RIFLE-TUBE READY FOR RECEIVING JACKET

with Professor Treadwell, of Harvard University, in 1841. He utilized it by enclosing a tube of cast-iron or steel in the same manner as it is applied in the wrought-iron Armstrong and Woolwich guns.

The administration of oar naval ordnance has abandoned conversions, and has concentrated its efforts on the production of an armament of built-up steel guns. The system of construction that has been adopted originated in England, but was for many years ignored by the government authorities. It involved the use of steel in all its parts, and this was charged as an objection, as confidence in this metal was not established in the minds of the English artillerists. That government committed itself entirely to the wrought-iron gun proposed by Mr. Armstrong, whose system was a reproduction of that successfully experimented on by Professor Treadwell, and the entire force of the government works at Woolwich and of the Armstrong works at Elswick-on-the-Tyne was occupied with the production of this style of ordnance. The English steel gun invented by Captain Blakely and Mr. J. Vavasseur was ignored in England, but its merit could not be suppressed, and its superiority has forced a tardy recognition by that government.

This gun came prominently into notice for a short time at the breaking out of the war of the rebellion: some guns were imported for the service of the Southern States. At the exhibition in London in

BREECH-LOADING RIFLE-JACKET, ROUGH-BORED AND TURNED

1863 a Blakely 8.5-inch gun was one of the features of attraction in the department of ordnance. The principle of the construction was shown in this gun, consisting in shrinking a long jacket of steel around an enclosed steel tube, the jacket extending to the trunnions. Mr. Vavasseur was the manager of the London Ordnance Works, and was associated with Captain Blakely in the manufacture of his earlier guns, but the entire business soon fell into the hands of Mr, Vavasseur, whose name alone is associated with the succeeding developments of the gun.

In 1862 the guns manufactured by Mr. Krupp were solid forgings. He advanced but slowly towards the construction of built-up cannons, and it was not until the failure of some of his solid-cast guns that he entered on the built-up system. His first steps were to strengthen the rear portion of new guns by shrinking on hoops, and to increase the strength of old guns he turned down the breech and shrunk on hoops. He confined this system of strengthening to the rear of the trunnions until he was reminded of the necessity of strength along the chase of the gun by the blowing off of the chase of some 11-inch guns of his manufacture. His system was then modified so as to involve reinforcing the tube of the larger calibred guns along its whole length with hoops, and his later and largest productions are provided with a long jacket reinforcing the entire breech portion of the tube—a virtual adoption of the great element of strength which has always formed the essential feature in the Vavasseur gun which is now adopted in the United States navy.

In the building up of the steel gun for the navy advantage is so taken of the elastic characteristic of the metal that all parts tend to mutual support. The gun proper consists of a steel tube and a steel jacket shrunk around it, reaching from the breech to and beyond the location of the trunnion-band. Outside the jacket and along the chase of the gun there are shrunk on such hoops as the known strain on the tube may make necessary for its support. The tube is formed from a casting which is forged, rough-bored, and turned, and then tempered in oil, by which its elasticity and tensile strength are much increased. It is then turned on the exterior, and adjusted to the jacket, the proper difference being allowed for shrinkage. The jacket, previously turned and tempered, is then heated, and rapidly lowered to its place. The front hoops over the chase are then put on, and the gun is put into a lathe and turned to receive the trunnion-band and rear and front hoops. The gun is then fine-bored and rifled.

Each part, as successively placed in position, is expected to compress the parts enclosed through the initial tension due to contraction

Putting the jacket on a 6-inch breech-loading rifle-tube

BREECH-LOADING RIFLE AFTER RECEIVING JACKET

in cooling. This tension is the greater the farther the part is removed from the tube; thus the jacket is shrunk on at a less tension than are the encircling hoops. By this means full use is made of the elastic capacity of the tube which contributes the first resistance to the expanding influence of the charge. The tension of the jacket prevents the tube being forced up to its elastic limit, and it in turn experiences the effect of the tension of the other encircling parts which contribute to the general support; thus no part is strained beyond its elastic limit, and on the cessation of the pressure all resume their normal form and dimensions. A comparison of this method of common and mutual support of parts with that given by the wall of a gun cast solid will serve to demonstrate the superior strength of the construction. In order to achieve this intimate working of all the parts it is necessary that the metal of which they are respectively composed must be possessed of the same essential characteristics; in a word, the gun must be homogeneous. It was the absence of this feature in the Armstrong gun which has caused its abolition. This gun was built up, and the parts were expected to contribute mutual support, but the want of homogeneity between the steel tube and the encircling hoops of wrought-iron made it impossible for them to work in accord, in consequence of the different elastic properties of the two metals, which, after frequent discharges, resulted in a separation of surfaces between the tube and hoops, when the tube cracked from want of support.

In the construction of the guns for the United States navy, as in the new steel guns now being manufactured in England, the theory of the built-up system is practically conformed to; more so than by Krupp or the French artillerists, who use a thicker tube than is considered judicious at Woolwich or at the Washington navy-yard. Any increase of thickness of the tube beyond what is necessary to receive the initial pressure of the charge is open to the objections made to the gun with a solid wall, the proportion of the strain communicated to the hoops is reduced, and rupture may ensue from overstraining the tube. The thicker the tube, the less appreciable must be the compression induced by the tension of the encircling hoops.

The gun is a breech-loader. The system adopted for closing the breech is an American invention, but having been employed in France

A Krupp hammer

from the earliest experimental period, it is known as the French *fermeture*. A screw is cut in the rear end of the jacket to the rear of the tube, and a corresponding screw is cut upon a breech-plug. The screw threads are stripped at three equidistant places, the screw and plane surfaces alternating, thus forming what is called an "interrupted" or "slotted" screw. The screw portions of the breech-plug enter freely along the plane longitudinal surfaces cut in the tube, and being then turned one-sixth of its circumference, the screw of the plug locks in that of the tube, and the breech is closed.

The success of this system of breech mechanism was not so pronounced on its introduction as it is today. The plug forms the base of the breech of the gun, and all the effort of the gases to blow out the breech is exerted at this point. The impact upon the end of the plug is very severe, and has a tendency to upset the metal, thereby increasing the diameter of the plug, which would prevent its removal after the discharge of the piece. With quick-burning powder, as was generally in use for cannons at the inception of the breech-loading experiments, this result ensued if the charges of powder were carried above a certain limit, and the consequent restriction that was put upon velocities was a serious obstacle to the adoption of the system; but the progress that has been made of late years in the science of gunpowder manufacture has relieved the subject from this embarrassment, powder being, now provided which communicates very high velocities while developing pressures so moderate and regular as to be entirely under the control of the artillerist. The original guns, four in number, constructed with breech mechanism on the French *fermeture* principle for the British government during the Crimean war are now in the "Graveyard" at Woolwich Arsenal.

The projectiles for the new armament are of two kinds; both, however, are shells. That for ordinary use against unarmoured vessels is styled the common shell, and is of cast-iron. The length bears a uniform proportion to the gun, being in all cases three and a half calibres. The armour-piercing shell is made of forged steel, and is three calibres in length. The following table gives the particulars, approximately, of the common shell:

Gun.	Length.		Weight.	Bursting Charge.
	Inches.	Calibres.	Pounds.	Pounds.
5-inch breech-loading rifle....	17.97	3.59	60	2
6-inch breech-loading rifle....	20.90	3.48	100	4
8-inch breech-loading rifle....	28.10	3.51	250	12
10-inch breech-loading rifle....	35.00	3.50	500	22
12-inch breech-loading rifle....	42.00	3.50	850	38
16-inch breech-loading rifle....	56.00	3.50	2000	90

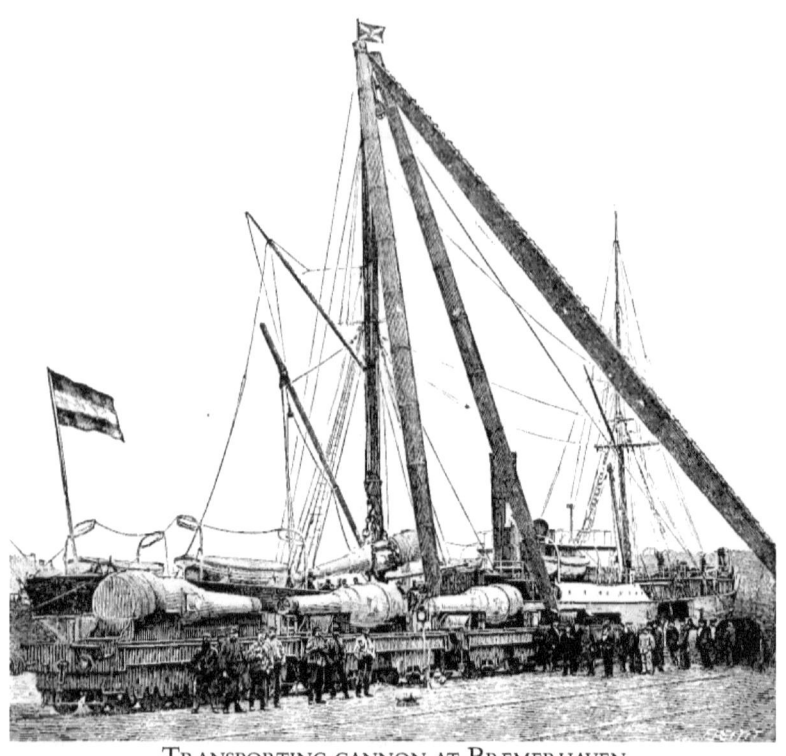
Transporting cannon at Bremerhaven

The armour-piercing shell of the same weight is reduced in length, and its walls are thicker; the bursting charge is consequently much reduced. The following are the particulars, approximately deter-mined:

Gun.	Length.		Weight.	Bursting Charge.
	Inches.	Calibre.	Pounds.	Pounds.
5-inch breech-loading rifle....	15.07	3.01	60	1
6-inch breech-loading rifle....	17.91	2.98	100	1.50
8-inch breech-loading rifle....	24.25	3.03	250	3.50
10-inch breech-loading rifle....	30.00	3.00	500	7
12-inch breech-loading rifle....	36.00	3.00	850	14
16-inch breech-loading rifle....	48.00	3.00	2000	30

The rifle motion is communicated by one rotating ring of copper, which is placed at the distance of 1.5 inch from the base of the projectile. The uniform windage for all calibres is .04 inch; thus, taking the 6-inch gun as an example, the diameter of the bore across the lands is 6 inches, the diameter of the shell is 5.96 inches, the depth of the grooves is .05 inch; thus the diameter of the bore across the grooves is 6.10 inches. In order to permit the rotating ring to fill the grooves, it must have a diameter of 6.14 inches; this causes a *squeeze* of .05 inch between the lands and the rotating ring.

There is no subject in the development of the new naval artillery more important than the powder. That used with the old artillery is entirely unsuited to the new conditions that obtain in the modern high-power guns. A brown powder, introduced first in Germany, has exhibited decided advantages over all others, and the efforts to reproduce it have been thoroughly successful at the Du Pont Mills. It is generally known as "cocoa" powder. Its peculiarity exists in the method of preparing the charcoal; this affects the colour, and results in a brown instead of a black powder. With this powder, experiments with the 6-inch gun give a muzzle velocity of over 2000 feet per second with a projectile of 100 pounds, using charges of 50 pounds, and this result is obtained with less than 15 tons pressure per square inch in the powder chamber. The grain is prismatic, with a central perforation, and as regards its rate of burning, is under complete control in the manufacture; the form provides an increasing surface for the flame during the period of combustion, thus relieving the gun from abnormal pressures at the moment of ignition, but continuing the extreme

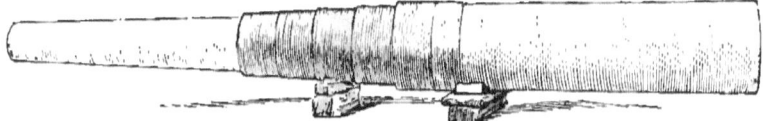

BREECH-LOADING RIFLE AFTER RECEIVING JACKET AND CHASE HOOPS

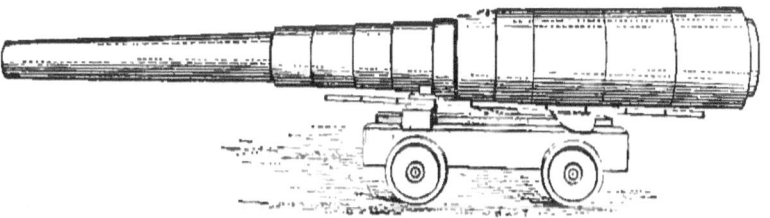

BREECH-LOADING RIFLE WITH JACKET, CHASE HOOPS, AND JACKET HOOPS IN PLACE

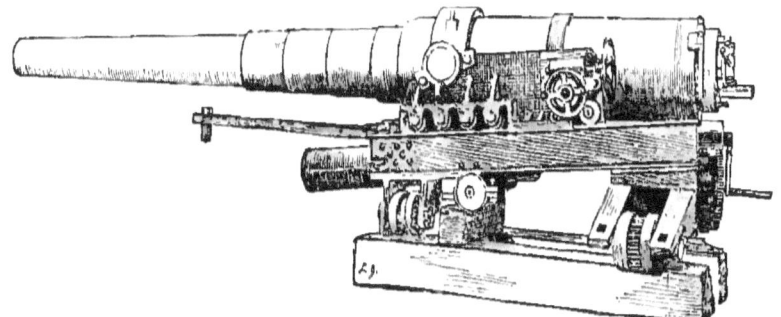

US. 6-INCH BREECH-LOADING RIFLE

pressure farther along the bore. The progressive nature of the combustion is very apparent when comparing an unburned grain with others partially consumed, blown out from the gun. The gun-carriage, which is a separate study in itself, is carried to a high pitch of perfection, and presents many features being adopted abroad. The importance of a suitable carriage can be appreciated by inspecting the following table, which exhibits the *energy* that must be controlled by it:

Gun.	Weight of Charge.	Weight of Projectile.	Muzzle Velocity.	Muzzle Energy.	Penetration in Wrought-iron.	Muzzle Energy per Ton of Gun.	Weight of Gun.	Weight of Carriage.
	Pounds.	Pounds.	Feet.	Ft.-Tons.	Inches.	Ft.-Tons.	Pounds.	Pounds.
5-inch steel breech-loading rifle..	30	60	1915	1,525	10.7	552	6,187	4,200
6-inch steel breech-loading rifle..	50	100	1915	2,542	13.2	524	11,000	6,400
8-inch steel breech-loading rifle..	125	250	2050	7,285	18.2	560	28,000	14,000
10-inch steel breech-loading rifle..	250	500	2100	15,285	23.7	588	55,240	32,482
12-inch steel breech-loading rifle..	425	850	2100	25,085	27.6	591	44 tons	....
14-inch steel breech-loading rifle..	675	1350	2100	41,270	32.2	550	75 tons	....
16-inch steel breech-loading rifle..	1000	2000	2100	61,114	36.8	571	107 tons	....

This *energy*, total energy, expresses the work that the gun can perform. It is expressed in foot-tons, and signifies that the energy developed is sufficient to raise the weight in tons to a height of one foot. Thus the projectile from the small 5-inch gun, weighing sixty pounds, fired with a charge of thirty pounds of powder, leaves the gun with an energy capable of lifting 1525 tons to the height of one foot! Comparing this with the energy developed by the 100-ton hammer at the forge of Le Creuzot in France, the energy of which is 1640 foot-tons, we have a most striking illustration of the power of gunpowder, and the testimony in the table as to the energy developed per ton of gun more forcibly exhibits the perfection of a manufacture which, with so little weight of gun, can develop such gigantic power.

It is this power, united with a moderate weight of gun, which will enable our unarmoured cruisers to hold their own with vessels moderately armoured. The power of the battery is greater than is required to contend with unarmoured ships, there is a great surplus of power of offence, and the effort is very properly made to sustain this at the highest practicable point. The table shows that the 5-inch gun can perforate 10.7 inches of wrought-iron at the muzzle; but the results given in tables are based on deliberate firing made on a practice-ground, with the position of the target normal to the line of fire. Such conditions cannot obtain during an action at sea, for, besides the modified effect caused by increased distance of target, it must be borne in mind that the side of an enemy's ship will be presented at varying angles, which introduces the element of deflection, than which no cause is more detrimental to penetration. Though the table states a fact, the practical effect of the projectile will be far less than is stated, hence the

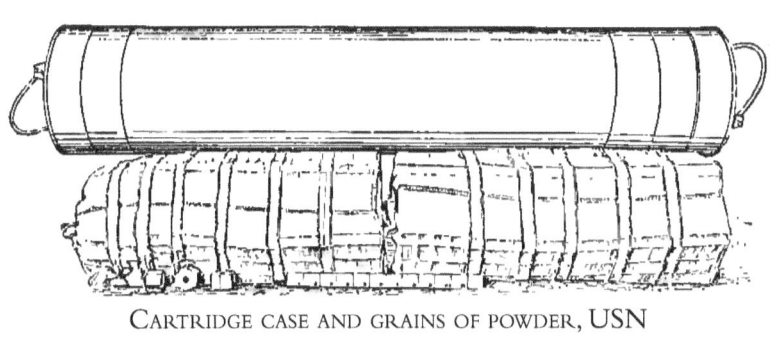

CARTRIDGE CASE AND GRAINS OF POWDER, USN

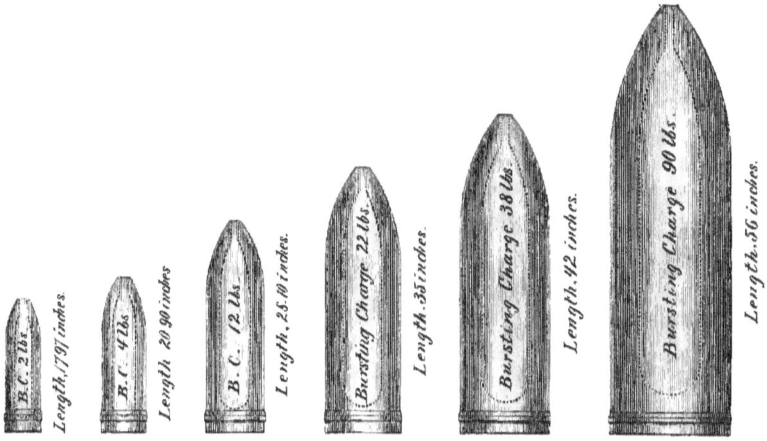

COMMON SHELLS, USN

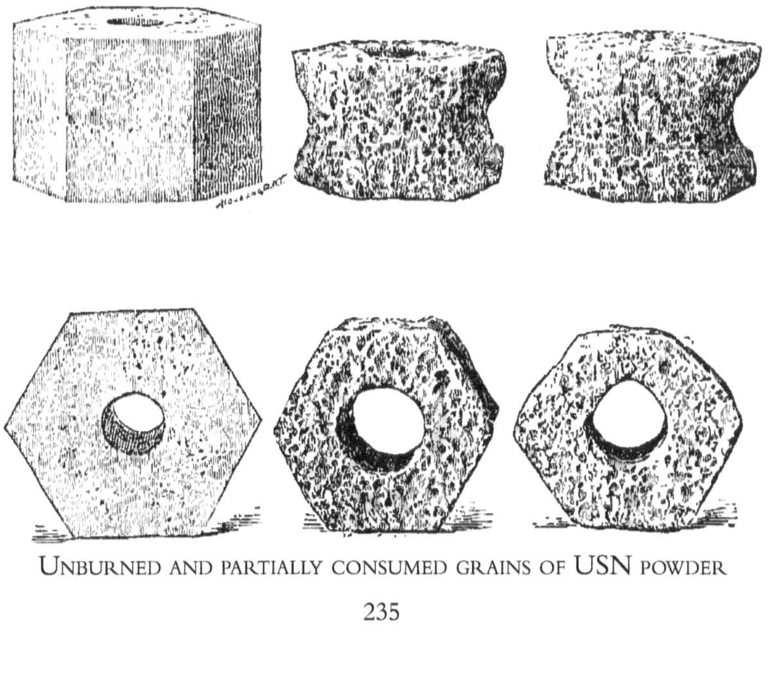

UNBURNED AND PARTIALLY CONSUMED GRAINS OF USN POWDER

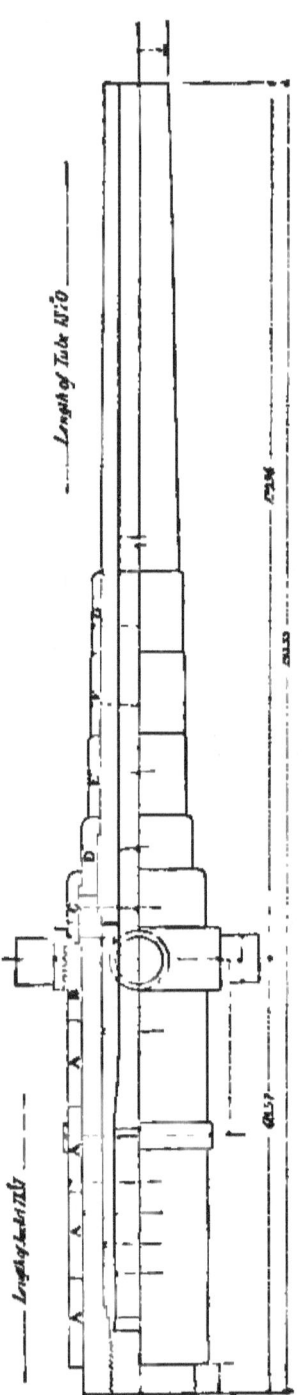

SECTION OF USN 6-INCH BUILT-UP STEEL BREECH-LOADING RIFLE

wisdom of providing a large surplus of power to compensate for the resistance to its operation.

It will readily be conceded that the artillerist has a very responsible duty to perform in so designing his gun that the parts shall lock and interlock to guard against chance of dislocation in the structure. A study of the illustration of the 6-inch built-up gun as constructed at the Washington navy-yard will show the system there adopted.

In the list of guns each calibre is represented by one gun. We have not, as of old, several guns of the same calibre differing in weight; multiplicity of classes will be avoided; but this will apply only to the main battery, for history is singularly repeating itself at this time in the restoration of the "murdering pieces" which have been cited as forming part of naval armaments in the seventeenth century. The needfulness of machine guns for operating against men on open decks, for effecting entrance through port-holes, for repelling attacks in boats, and for resisting the approach of torpedo-boats, is so widely recognized that no vessel of war is considered properly equipped without a secondary battery of these "murdering pieces." They are mounted on the rail, on platforms projecting from the sides and in the tops. The types adopted in the United States navy are the Hotchkiss revolving cannon and rapid-firing single-shot guns, and the smaller calibre machine guns of Gatling. The heavier pieces, throwing shells of six pounds weight, are very effective against vessels of ordinary scantling.

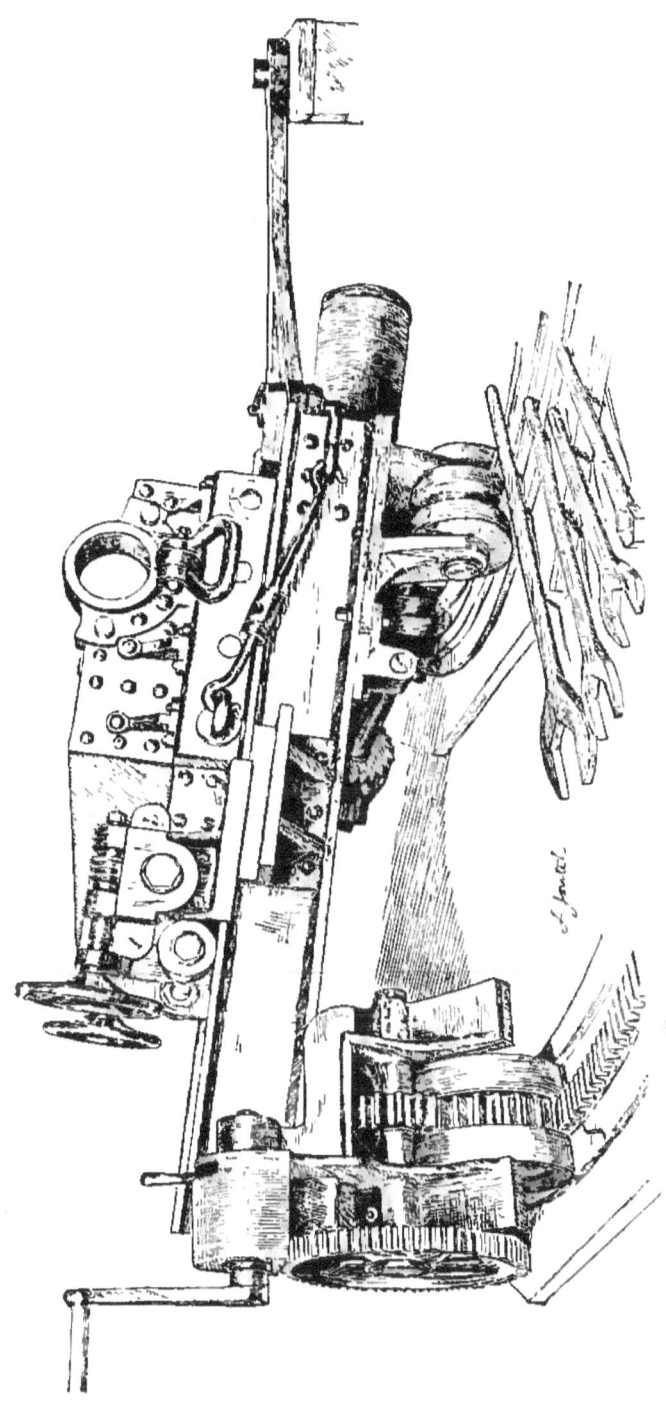

Broadside carriage for 6-inch breech-loading rifle

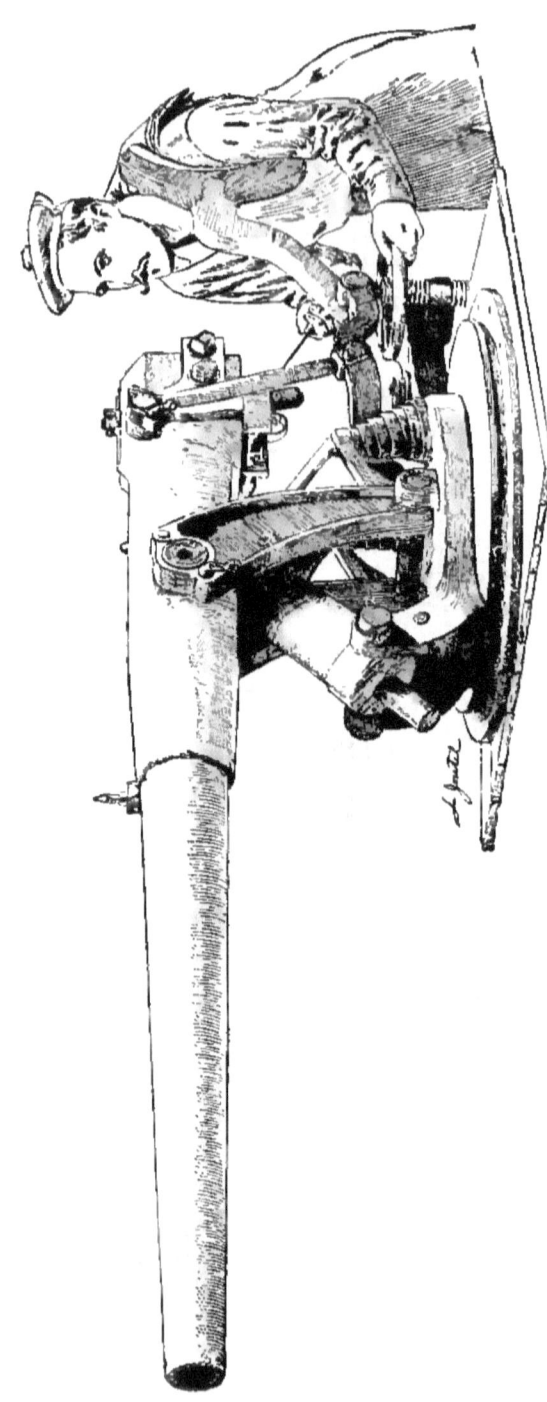

RAPID-FIRING SINGLE-SHOT HOTCHKISS GUN

In contemplating the present condition of our new naval armament we have the consolation of knowing that, so far as concerns the study of the subject generally and in detail, the designs, and the initial manufacture, all has been done that could have been done with the resources available. What has been achieved has been without the facilities that are provided in modern gun factories; but notwithstanding all the drawbacks, it is probably safe to assert that no guns in the world today are superior to those that have been fabricated at the Washington navy-yard of steel on the new adopted pattern. The work at this ordnance yard is carried on without ostentation; there is no flourish of trumpets accompanying its operations; it is not advertised, and the people do not yet know how much they owe to the ordnance officers of the navy for the initiation of this new industry, which enables us to assert our ability to advance in this manufacture through the incontrovertible proof of work accomplished. The results are meagre in quantity, and at the present rate of manufacture it will require many years to equip our fleet with modern artillery; this should be remedied, as there is now no doubt as to the success of the productions of this establishment. The plant should be enlarged on a liberal and well-matured plan, and the work should be encouraged by generous appropriations.

It may not be generally known that the steel forgings required for the few 8-inch and the two 10-inch guns now in hand were imported from abroad, for the reason that they could not be furnished of domestic manufacture, from the want of casting and forging facilities in the United States for work of such magnitude. This was a deficiency in our resources that required prompt attention to secure us a position of independence in this important matter. The method of achieving the object was carefully studied out by a mixed board of army and navy officers, and presented in a document known as the "Gun Foundry Board Report," and the subject received the attention of committees from both Houses of Congress. All of these reports virtually agreed as to the method, but there was a useless delay in action; large expenditures of money were required, and there was hesitancy in assuming the responsibility of recommending it. The object was of national importance, however, and public opinion demanded its accomplishment. The officers of the navy have proved their ability to carry on the work successfully; and if the opportunity be given they will establish the artillery of the United States navy in a position of which the country may again be proud.

Notes

Guns

The United States no longer depend upon foreigners for guns or armour, inasmuch as the circular issued in August, 1886, by the Navy Department inviting all domestic steel manufacturers to state the terms upon which they were willing to produce the steel plates and forgings required for ships and ordnance, has met with a prompt response. About 4500 tons were needed for armour, in plates varying from 20 feet by 8 feet by 12 inches thick, to 11.6 feet by 4.3 feet by 6 inches thick; and of the 1310 tons of steel forgings, 328 tons were intended for the 6-inch guns, 70 tons for the 8-inch, and 912 tons for the calibres between 10 and 12 inches, both inclusive. The rough-bored and turned forgings required by the contract were to weigh 3¼ tons for the 6-inch calibres, 5 tons for the 8-inch, 9½ tons for the 10-inch, 9¾ tons for the 10½-inch, and 12½ tons for the 12-inch. From the time of closing the contract twenty-eight 6-inch forgings were to be delivered in one year, and the remainder within eighteen months. All the 8-inch were to be ready within two years, and the 10-inch and larger calibres within two years and a half. The proposals opened on the 22nd of last March showed that for the gun-forgings the Cambria Iron Company had bid $851,513, the Midvale Steel Company $1,397,240, and the Bethlehem Iron Company $902,230; and that for the armour-plates the Bethlehem Company had bid $3,610,707, and the Cleveland Rolling-mill Company $4,021,561. Subsequently the Navy Department awarded the contract to the Bethlehem Company, which agreed to furnish all the required steel at a total cost of $4,512,938.29.

The tests are so rigorous that a high quality of steel is sure to be produced. The specifications require the forgings to be of open-hearth steel of domestic manufacture, from the best quality of raw material, uniform in quality throughout the mass of each forging and throughout the whole order for forgings of the same calibre, and free from slag, seams, cracks, cavities, flaws, blow-holes, unsoundness, foreign substances, and all other defects affecting their resistance and value. While it is prescribed that the ingots shall be cast solid, latitude is given to the method of production; but no matter what method may be employed, the part to be delivered for test and acceptance must be equal in quality and in all other respects to a gun ingot cast solid in the usual way, from which at least 30 *per cent*, of the weight of the ingot has been discarded from the upper end and 5 *per cent*, from the lower end.

For breech-pieces each ingot must be reduced in diameter by forging at least 40 *per cent.*; in case tubes are forged upon a mandrel from bored ingots, the walls must be reduced in thickness by forging at least 50 *per cent*. Forgings are to be annealed, oil tempered under such conditions as will assure their resistance and again annealed, and no piece will be accepted unless the last process has been an annealing one. The forging must be left with a uniformly fine grain.

All these excellent results are the direct outcomes of the report made in 1884 by the Ordnance Board. 1st. That the army and navy should each have its own gun-factory; 2nd. That the parts should be shipped by the steel-makers ready for finishing and assembling in guns; 3rd. That the government should not undertake the production of steel of its own accord; 4th. That the Watervliet Arsenal, West Troy, N. Y., should be the site of the army gun-factory; and 5th. That the "Washington navy-yard should be the site of the navy gun-factory. No action was taken upon the recommendation to establish gun-factories; but at the first session of the Forty-ninth Congress an appropriation of $1,000,000 was made for the armament of the navy, of which sum so much as the secretary determined might be employed for the creation of a plant. Under this permission the gun-factory at the Washington navy-yard is now being established.

The construction of the breech-loading steel guns for the new cruisers has been energetically pushed. Slight modifications in the original designs were made necessary by the adoption of slower burning powder, which carried the pressure still farther forward in the bore, and, in the case of some foreign guns, caused their destruction. Though our guns have not suffered from any such accident, it has been deemed a wise precaution to give the 8-inch guns of the *Atlanta* two additional chase-hoops, and to hoop all other pieces of this calibre to the muzzle.

From a memorandum kindly furnished by Lieutenant Bradbury, United States navy, it is learned that the number and calibre of the new guns now finished, under construction, or projected, are as follows:

Name of Ship.	Calibre.				
	5-inch.	6-inch.	8-inch.	10-inch.	12-inch.
Dolphin	None.	1	None.	None.	None.
Atlanta	"	6	2	"	"
Boston	"	6	2	"	"
Chicago	2	8	4	"	"
Gun-boat No. 1	None.	6	None.	"	"
Gun-boat No. 2	"	4	"	"	"
Newark	"	12	"	"	"

Name of Ship	Calibre.				
Baltimore	"	6	2	"	"
Charleston	"	6	None.	2	"
Miantonomoh	"	None.	"	4	"
Terror	"	"	"	4	"
Amphitrite	"	"	"	4	"
Monadnock	"	"	"	4	"
Puritan	"	"	"	4	"
Armored cruiser	"	6	"	4	"
Armored battle-ship	"	6	"	None.	2
2 Gun-boats	"	12	"	"	None.
2 Cruisers	"	24⅔	"	"	"
Floating batteries	"	None.	"	"	8

This gives a total of two 5-inch, one hundred and three 6-inch, ten 8-inch, twenty-six 10-inch, and ten 12-inch. In his last report. Captain Sicard, Chief of Ordnance, states that:

> For the new ships approaching completion we have eighteen 6-inch, three 8-inch, and two 5-inch guns finished, and three 6-inch and five 8-inch well advanced, together with all the carriages for the *Atlanta* and *Boston*, and all for the *Chicago*, except the 8-inch. ...With brown powder the following are the best results obtained in the 6-inch and 8-inch guns. It will be observed that the muzzle velocities are as high, while the chamber pressures are considerably below those which the guns were calculated to support in service.

Gun.	Powder.	Muzzle Velocity.	Pressure.
		Foot seconds.	Tons.
6-inch	American Brown.	2,105	15.6
8-inch	Westphalian Brown.	2,013	15.5

During the preliminary trials afloat of the *Atlanta*'s battery in July, a few minor faults were unfairly given an importance by the newspapers which led the country to believe that the ship and her armament were useless. Unfriendly critics vented their spite and aired their ignorance in condemnations which included all who had had anything to do, even in the remotest degree, with the design and construction of vessel and gun. Indeed, so bitter and persistent were they that for a time it seemed almost hopeless to expect any further good could come out of the Nazareth of public opinion. It was not a question of politics, for the journalists of every political faith ran amuck riotously upon the subject; nor was it a matter of morals, where, through intelligent discussion, better things could be attained, for with brilliant misinformation and dogmatic dullness each scribe stuck his pin-feathered goose-quill into the navy's midriff—it being such an easy, such a safe thing to do—and then thanked Heaven he was a virtuous citizen. Finally, a board was appointed to inspect the ship and battery, and after a thorough examination it made the following report:

> In obedience to the Department's order of the 22nd instant, the Board convened on board the *Atlanta*, Newport, Rhode Island, on the 25th instant (July, 1887), and made a careful examination of the ship, guns, carriages, and fittings, and of the damage sustained during the recent target practice, as reported by the board of officers ordered by the commanding officer of the *Atlanta*. The Board proceeded to sea on the morning of the 25th instant, but were prevented from firing the guns by a heavy fog which prevailed throughout the day. The ship was again taken to sea on the morning of the 27th instant, and the guns were fired. No deficiencies were noted in the guns themselves other than a slight sticking of the breech-plug in 6-inch breech-loading rifle No. 5 (this disappeared during the firing), some difficulty in the management of the lock of 6-inch breech-loading rifle No. 4, caused by slight upsetting of the firing-pin, and the bending of the extractor in 6-pounder rapid-fire No. 5.
>
> The recoil and counter-recoil of the 8 and 6 inch guns were easy and satisfactory, except at the second fire of the 8-inch breech-loading rifle No. 1, when the gun remained in. (This was readily run out with a tackle.) The action of the carriage of 8-inch breech-loading rifle No. 1 at the first fire was due to want of strength in the clips and clip circles, and at the second fire to want of sufficient bearing and securing of the deck socket. It is believed that had the deck socket held, the carriage would not have been disabled by the giving way of the clips. The training gear, steam and hand, was uninjured; the gun was readily trained when run out to place. The action of the after 6-inch shifting gun No. 4 was satisfactory, notwithstanding that the front clips had a play of half an inch. The action of the broadside carriages of 6-inch guns Nos. 5 and 18 was satisfactory, except the breaking of clips, the starting of the copper rivets in the clip circles, and the wood screws in the training circles.
>
> It is believed from the action of the carriage of 6-inch breech-loading rifle No. 5, when the clips were removed, that the carriages can be safely used without clips. The clips, however, give additional security and steadiness to the carriage, and assist the pivot and socket in bearing the shock of the discharge.

The firing of the 6-pounder rapid-fire guns developed a weakness in one leg of the cage mount of No. 4, due to imperfect workmanship, and showed also the necessity of locking nuts on the bolts that secure the mounts to the ports. The tower mounts of the 3-pounder rapid-fire guns are unsatisfactory. They cannot be moved with facility; the line of sight of the gun is obstructed at ranges beyond 1600 yards, and the guns cannot be safely used as now fitted. For this reason 3-pounder rapid-fire No. 3 was not fired. The tripod mounts of the 1-pounder rapid-fire guns need stronger holding-down arrangements. The tower mounts of the 47-millimetre revolving cannon are like those of the 3-pounder rapid-fire guns, and have the same defects. The mounts of the 37-millimetre in the tops are satisfactory.

Careful observation of the effect of the firing upon the hull of the vessel failed to develop any damage other than the breaking of the cast-steel port-sills and the starting of some light wood-work. The shock of discharge was slight on the berth-deck, and observers there were unable to observe which 6-inch gun had been fired. The deck, hull, and fittings, with the exception of the port-sills, hinges to superstructure doors and vegetable lockers, and some of the light wood-work, have every appearance of strength and ability to endure the strain of continuous firing of the guns. The blast of the forward 8-inch gun, when fired abaft the star-board beam, will not permit the crews of the starboard 3-pounder rapid fire and 1-pounder rapid fire to remain at their guns. When the after 8-inch gun is fired forward of the port beam, the crews of the after 47-millimetre revolving, cannon and of the port after 1-pounder rapid fire cannot remain at their guns. When the forward 6-inch shifting gun is fired on the port bow or directly ahead, the crew of forward 8-inch gun cannot remain at their places. When the after 6-inch shifting gun is fired on the starboard quarter or directly aft, the crew of the after 8-inch gun cannot remain at their gun. The inability to fire parts of the secondary battery under certain conditions is due to the great arc of fire given to the 8-inch guns. This can hardly be called a defect. It is thought that a screen can be placed between the 8 and 6 inch guns which will enable them to be worked together forward or aft.

The pivot socket of the 8-inch carriage should have a broader bearing surface, and should be rigidly bolted to the steel deck and to the framework of the ship in such manner as to distribute the strain over a larger area. The clips and clip circles of the 8-inch and 6-inch carriage should be made of steel. The clips should have larger bearing surfaces, and should be shaped to fit the circle. The circle should have double flanges, and be bolted (not riveted) on each flange to the steel deck. There should be no appreciable play between the clips and the circles. All bolts used in the battery fittings should have the nuts locked.

The clip rail of the tower mount should be altered to fit the mount. This change will make the compressors effective, and allow the guns to be used with safety. The port-sills should be replaced by heavier sills, made of the best quality of malleable cast-steel. The plan of testing the hull, guns, and fittings of the *Atlanta* arranged by the Board contemplated a more extended use of the main battery, but the weakness developed in the port-sills and in the sockets of the 8-inch carriages rendered further firing inadvisable.

Whatever conclusion may be drawn from this report, there is one fact which may serve as an important corollary. In the latest drills of the ships on the North Atlantic station, the *Atlanta* won the champion pennant for the best gunnery practice, and this with guns and carriages which were said to be completely disabled.

The safe employment of high explosives for war purposes is looked upon by many as a solution of certain vexed problems, and much time and money have been given to the subject. From the nitro-glycerine products there has been a loudly heralded advance to melinite and roborite, of which the great things expected have not yet been realized. Among the most promising attempts to use dynamite in a projectile is that made with the pneumatic gun, perfected by Lieutenant Zalinski, of the U. S. Artillery, who has courteously furnished the following description of the system:

> The pneumatic dynamite torpedo gun is a weapon which has been evolved for the purpose of projecting with safety and accuracy very large charges of the high explosives. While a gun in name and form, it is practically a torpedo-projecting machine, the propelling force used being compressed air. The use of the compressed air gives uniformity and complete control of pressures and total absence of heat. This insures entire absence of violent initial shocks from the propelling force; it also eliminates danger of increasing the normal sensitiveness of the high explosives by heating while resting in the bore of the gun.

The ability to reproduce, time after time, absolutely the same pressure necessarily carries with it great accuracy of fire. The torpedo shell thrown by the gun is essentially arrow-like, and is very light and compact compared to the weight of charge thrown. This is a matter of no little importance on shipboard, as a very much larger number can therefore be carried for a given weight and storage room. The torpedoes projected by this machine have a twofold field of action when acting against ships: first, the over-water hull, second, the under-water hull.

The shell is exploded by an electrical fuse. This is brought into action if striking the over-water hull an instant *before* full impact. If the shell misses the over-water hull and enters the water, explosion is produced *after* the shell is thoroughly buried, thus obtaining the fullest tamping effect of the water. The delayed action of the fuse can be controlled so as to cause the shell to go to the bottom before explosion ensues. This is needed at times when the torpedo shell is used for counter-mining a system of submerged stationary torpedo defences.

Experiments against iron plates have shown that it is essential to have the initial point of explosion at the rear of the shell. When explosion takes place by simple impact from the front end, the injury to the plates is actually less than when a blank shell is used.

For these reasons the fuse has been arranged so that the initial point of explosion is at the *rear* of the shell. No attempt has been made to make a shell which can perforate armour before explosion. To do so would involve thickening the walls to such an extent as to materially reduce the weight of the charge carried. Besides that, it is very doubtful whether a shell fully charged with gunpowder can perforate any considerable thickness of armour without previously exploding its bursting charge. Much more will this be the case where the bursting charge is one of the more sensitive high explosives.

The pneumatic torpedo-gun system has various fields of usefulness as an auxiliary war appliance. Among these are the following:

1st. On swift-moving torpedo-boats;

2nd. On larger war-vessels, for general use and for defence against surface and submarine torpedo-boats;

3rd. In land defences;

4th. For use in the approaches during land sieges.

Torpedo-boats carrying the pneumatic guns can commence effective operations at the range of at least one mile, as compared to not more than three hundred yards of the boats carrying the Whitehead torpedoes. Their torpedo shell cannot be stopped by netting, as is the case with the latter. The charges which can be thrown are also much greater. The guns to be carried on the pneumatic dynamite gun cruiser now building for the United States government will throw shell charged with 200 and 400 pounds of explosive gelatine. These guns can be fired at the rate of one in two minutes, and indeed even more rapidly if required.

In the defence of a man-of-war no other means can as effectually stop the advance either of submarine boats or submerged movable torpedoes. This is due to the ability to explode the large charges when the shells are well submerged. Their radius of action will be so great as to avoid the necessity of making absolute hits. The chances of stopping the attack are thereby very much increased.

A tube of large calibre can be fixed in the bow, so as to be of use when advancing to the attack with the ram. An 18-inch shell, containing 1000 pounds of explosive gelatine, can be thrown 500 yards in advance of the ship, and that, too, without danger of running into the explosion of its own petard, as would be the case in ejecting directly ahead ordinary torpedoes. This will be made more clear by the statement of the relative speed of the two classes. The pneumatic gun torpedo has a mean velocity of 400 knots for a range of one mile, as compared to 25 knots for a range of 200 yards of the Whitehead torpedo. Furthermore, there is no danger of the shell turning back, as is sometimes the case with the latter.

The opportunities of making an effective hit will be much greater with the torpedo shell than with the ram; it will be easier to point the vessel fairly at the enemy's broadside when at the range of five hundred yards than to bring the ram in absolute contact with the enemy's side. The gun-tubes used are very thin (not exceeding three-quarters of an inch in thickness), and may be of sections of any convenient length. The other portions of the supporting truss, reservoirs, etc., are also of comparatively light weight. They could be of large calibres, and the destructive effects producible by large charges of high explosives will doubtless have a demoralizing effect upon the defence.

Upon September 20th of this year a public trial was successfully made with the gun, the target being the condemned coast survey schooner *Silliman*. After firing two shots to verify the range, the gun was loaded with a projectile which was five and a half feet in length, contained fifty-five pounds of explosive gelatine, and was fired under an air pressure of 607 pounds. The torpedo rushed from the muzzle of the tube with a loud report; in thirteen seconds it plunged into the water close under the starboard quarter of the *Silliman*, and exploding almost instantly, threw a great volume of water one hundred and fifty feet into the air.

For a moment the schooner was hidden from view, but when the mist cleared away it was found that her main-mast had toppled over the side. At a distance this seemed to be all the damage inflicted, but a closer inspection showed that all the wood-ends on deck had been loosened, that the cabin fittings had been thoroughly shaken up, and that water was running into the hold.

Soon afterwards a fourth shot was fired. This landed very close to the starboard side of the vessel, and on explosion seemed to lift the *Silliman* out of the water.

The hull was very badly shattered; the water-tank, which had been firmly fastened to the schooner's bottom, was blown up through the deck and floated on the wreckage, and the stump of the main-mast was capsized. The bow was held above water by barrel buoys, and the fore-mast, which had heeled over to an angle of forty-five degrees, was sustained by the steel rigging that had become entangled in the pieces of wood floating to windward.

Machine and Rapid-Fire Guns

Or the machine guns, the Gatling, Gardner, Nordenfeldt, and Maxim systems are the best known. The adoption of the Accles feed in the Gatling eliminates largely the liability of cartridge jams, and increases the rapidity of fire at all angles to twelve hundred shots per minute; when this rapid delivery of fire is not needed, Bruce's slower feed may be substituted. The Gardner gun is an effective weapon, but it has less rapidity of fire and smaller range of vertical train than the Gatling. The Nordenfeldt rifle-calibre gun has not obtained the prominence of the others, and the Maxim, in which the energy of recoil is ingeniously applied to the work of loading and firing, is growing in favour. The Hotchkiss revolving cannon was a wonderful step— the 37, 47, and 53 millimetre calibres firing 1 pound, 2½ pound, and 3½ pound explosive projectiles, with muzzle velocities of about 1400 feet per second. Commander Folger, United States navy, declares:

> The heavier nature of revolving cannon proved somewhat unwieldy, and the change to the single barrel of increased length, and using a heavier powder charge, was a natural one, and in keeping with the growing ballistic power of large guns. Though no longer denominated machine guns, the term now being generally applied to a cluster of barrels, the rapid-fire guns are-a direct outgrowth of the larger calibres of machine guns, and are classed with them as secondary battery arms. There are now in the service of all the great military powers rapid-firing guns of 47 and 57 millimetre calibre, firing respectively explosive shells of 3 pounds and 6 pounds weight, at muzzle velocity of about 1900 feet per second. This will give with the 6-pound gun a range of about 2½ miles at 10 degrees elevation. These guns will deliver, under favourable circumstances, perhaps ten aimed rounds per minute, and the shells perforating the sides of an unarmoured vessel, and bursting, after passing through into, say, twenty-five fragments, each with energy sufficient to kill a man, we have here a weapon of unequalled destructive capacity. It is beyond question that the conditions of combat between ships and forts are definitely changed by the advent of these guns. Even armoured vessels with covered batteries are at a disadvantage, as a hail of missiles will seek the gun-ports and conning-towers wherever an enemy, from the nature of circumstances, takes close quarters. Experiment abroad has also demonstrated that the projecting chase (forward body) of a large gun is extremely vulnerable, and liable to injury from the fire of the larger rapid-firing pieces.
>
> This system, which is just now so important an adjunct to the main battery of ships of war, is of but recent development. The first order received for a weapon of this kind by the Hotchkiss firm came from the United States, and the guns now mounted in the new ships *Boston, Atlanta,* and *Dolphin* were delivered under it. Three calibres were obtained, *viz.*, the 6, 3, and 1 pounder, as they are known in the United States navy, their usual names in other countries being the 57, 47, and 37 millimetre guns. Since their introduction the demand for larger calibres by most of the prominent naval powers has been so pressing that the Hotchkiss Company has produced a 9-pounder and has a 33-pounder in course of manufacture. It is

believed that this last calibre represents about the limit of utility of the Hotchkiss system, though the gain in time by the use of ammunition carrying the charge projectile and fulminate in one case will recommend it for use with much larger calibres, even where two men may be required to handle the cartridge.

The most important trials of rapid-fire guns during the past two years are thus described by Lieutenant Driggs, United States navy:

The various systems now in use, or being developed, are the Albini, Armstrong, Driggs-Schroeder, Gruson, Hotchkiss, Krupp, Maxim, Nordenfeldt. Of these the Armstrong has not been favourably received on account of the cumbersome breech-closing arrangement. This consists in two side levers attached to and turning about the trunnions; a cross-head connects the two levers, and by an eccentric motion one of them is pressed against or removed from the breech of the gun, thus closing or opening it. The *Bausan* has two of those guns, but with that known exception few, if any, have been put in service,

The Gruson gun is said to be very similar to the Hotchkiss in its mechanism, though not as good. The Maxim and Hotchkiss are both well known. The Nordenfeldt, which in Europe is the greatest rival of the Hotchkiss, is entirely different from the guns heretofore made under this name. In the single-shot rapid-fire gun the breech is closed with a double breech-plug, which is revolved in the breech recess by a cam motion. The plug is divided transversely; the front half carries the firing-pin, and has only a circular motion in closing and opening; the rear half acts as a wedge, the first motion being downward and the second circular, the front half then moving with it.

One of the most complete tests to which guns of this class have been subjected was that conducted by the Italian government in February of last year (1885). The trials were made-at Spezia, the following being offered for test: Hotchkiss rapid-fire; improved Nordenfeldt rapid-fire on recoil carriage; Armstrong rapid-tire; and a rapid-fire gun made at the Government Works at Venice. The Armstrong gun was not fired; the others were fired in the following order: Nordenfeldt, Hotchkiss, and Italian.

The guns were mounted on board a small ship (the *Vulcano*) for firing at sea. A large target was fixed on the breakwater in the middle of the harbour of Spezia, and two smaller targets of triangular shape had been anchored, one 550 yards inside, and the other 550 yards out-side, the breakwater. The *Vulcano* was then placed 1300 to 1400 yards inside the breakwater, and fire begun against the large target with the Nordenfeldt 6-pounder gun, which was worked by Italian sailors. A first series of eighteen shots were fired in forty-seven seconds, for rapidity of fire with rough aiming. A second series of sixteen shots were fired in thirty-four seconds. The rapidity of fire with rough aiming and untried men was thus respectively at the rate of twenty-three and twenty-eight shots per minute. Afterwards, ten case-shots were fired with the gun almost level, in order to see how the lead bullets were spread over the range. Some of them were seen to touch the water 700 or 800 yards from the muzzle, and the whole range was well covered by the 150 lead bullets contained in each of the Nordenfeldt case-shots.

The second part of the programme consisted of the firing at three targets, respectively at 600, 1200, and 1800 yards, the ranges being only approximately known, changing the aim at every third shot, and firing under difficulty, owing to the movement of the ship. Twenty-one common shells were fired, seven at each target, with good accuracy, and the shells on striking the water burst better at the shorter than at the longer ranges.

The firing at sea was closed with one more series of ten rounds, fired rapidly in twenty-six seconds, in order to see if the gun would act well after being heated by the eighty-five rounds which had already been fired. Four of the last series were ring-shells, and burst on striking the water at the first impact, breaking into a larger number of pieces than the common shells. The Nordenfeldt gun was then mounted on shore for tests of penetration. The plates used were:

1. A 5¼-inch solid wrought-iron
2. A 4-inch solid (Cammell) steel plate
3. One $7/_8$-inch steel plate, at an angle of fifteen degrees to line of fire. The two thick plates were backed by ten inches of oak, and at right angles to line of fire and one hundred yards from the gun. The perforation was in every case complete, both with solid steel shot and chilled-point shells, these latter bursting in the wood behind. The thin plate was then put at more acute angle to the line of fire, and only when this angle was seven degrees or eight degrees did the projectile fail to penetrate. The indicated muzzle velocity of this gun is 2130 feet, with a 6-pound projectile and charge of two pounds fifteen ounces.

A few days afterwards the Hotchkiss gun went through the same trials and programme. For rapidity forty rounds were fired with rough aiming in three minutes, the rate being 13.3 per minute. The shooting was good, but the men serving the gun complained of being fatigued by the shocks from the shoulder-piece. The muzzle velocity was about 1085 feet, or about 300 feet below that of the Nordenfeldt. Last of all, the Italian gun was fired, but as it was designed for 1480 feet velocity, it was not brought in direct competition with the other two guns in power. The rapidity of fire obtained, however, was about twenty rounds per minute, and both the mechanism and recoil-carriage worked well.

The Hotchkiss and Nordenfeldt guns were tried in competition at Ochta, near St. Petersburg, in September last (1886). The reports that have reached this country are very meagre, but are unanimous in favour of the Nordenfeldt gun. From what can be learned, the fire was first for rapidity, in which the Nordenfeldt discharged thirty rounds in one minute, and the Hotchkiss twenty rounds in the same time, the initial velocity of the former being 624 metres (2047 feet) per second, while that of the latter was 548 metres (1797 feet) per second.

The fire of both guns was directed upon a target at 1800 metres (1969 yards) range. The Nordenfeldt scored nine hits, while the Hotchkiss made none. It is more than likely that this failure was due more to defective pointing than to any defect of construction. The trial closed with a very interesting and instructive experiment.

Four targets were placed at 600, 800, 1000, and 1200 metres; each gun was to fire as rapidly as possible for thirty seconds, changing the range each fire, from the 600 up to the 1200 metre target and back. During this test the Nordenfeldt is said to have discharged fifteen shots in the thirty seconds, and to have made nine hits, while the Hotchkiss scored but two hits and only discharged eleven rounds in thirty-two seconds. Here again the element of inaccurate sighting may be largely responsible for the difference in the number of hits, but the great disparity in the number of rounds fired must be due to the mechanical defects in the Hotchkiss system by which the action of its breech-block is too slow. Notwithstanding the reported success of the Nordenfeldt gun in the trials, the Russian government ordered a number of Hotchkiss guns and no Nordenfeldts.

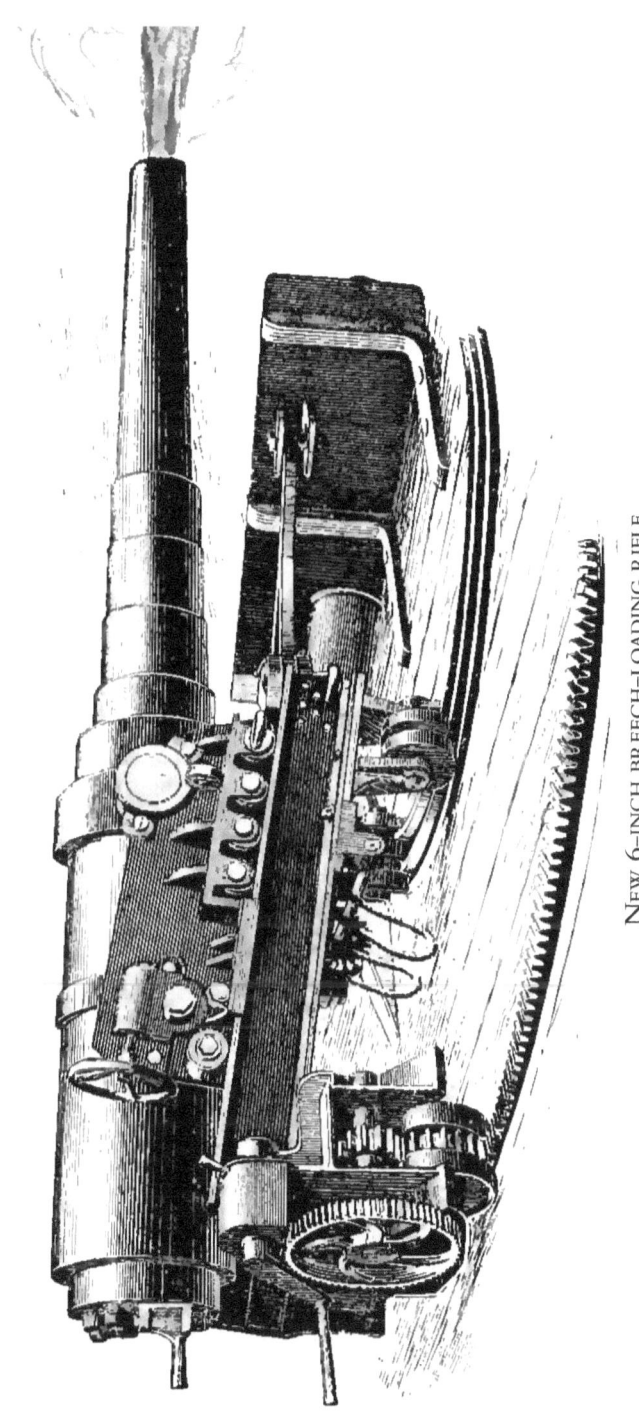

New 6-inch breech-loading rifle

The latest experiments with large calibred rapid-fire guns were those of the Armstrong 36 and 70 pounder. The first piece differs materially from the new 33-pounder Hotchkiss; it is 4.724 inches in calibre, 14 feet 2½ inches length, and weighs 34 hundred-weight. It was fired with seven and a half pounds of powder ten times in forty-seven seconds, or at a rate six times faster than that obtained with the service guns of like calibre. The 70-pounder was fired with both twenty-five-pound and thirty-pound charges, at a speed of from eight to ten rounds per minute. In the latest mount for the 36-pounder the gun is supported on a rocking slide which pivots on transverse bearings, so that the piece moves only forward and backward on the slide; elevation and depression are given by a shoulder-piece attached to the slide, and the gun is secured at any desired angle with a clamp attached to the side of the slide.

This development of rapid-fire pieces opens anew the discussion as to the comparative values of large and small calibre guns. At the present stage of the question it is safe to say that, however necessary the large calibre may be in armoured battleships and coast-defence vessels, its usefulness in thin-skinned, high-powered cruisers is questionable. Abroad, the long-range guns which constitute the primary batteries are being reduced in calibre, while the secondary batteries of rapid-fire guns are increasing so much in size that before the next sea-war a nearly uniform calibre of four or five inches will probably be established.

The reasons for these changes are not difficult to understand. In all sea engagements hereafter type will fight with type; that is to say, apart from the role which auxiliary rams and torpedo-boats may play, armoured ships will oppose armoured ships, and unarmoured cruisers and gun-boats will, when intelligently handled, seek action only with vessels of similar character. Today, (as at time of first publication), every unarmoured ship afloat or under construction can be penetrated at the average fighting distance by a musket-bullet impelled with a little more than the ordinary velocity; and as there is absolutely no protection, it seems a mistake to arm such vessels with the unnecessarily large calibres now in use. Especially is this true when their employment is based mainly upon the remote assumption that such ships may have to attack fortifications. Smaller guns will do the work equally as well, if not better; for the greater intensity of fire secured by the certain action of a large number of easily handled small-calibred guns is surely more valuable than any probable advantage which might be derived from heavier projectiles fired under conditions that make their effectiveness doubtful.

Whatever may be said to the contrary by mere theorists, the difficulty of handling ordnance increases enormously as the calibres grow; and sea-officers, who alone are the proper judges, insist that the monster pieces of the present day are so unmanageable as to be nearly useless, Of course, where armour penetration is vital to success, heavy armaments must and will be employed; but when this factor need not be considered, a great many light guns, easily worked by hand, are the demands of the hour. The problem, fortunately, is nearer solution owing to the development now in progress; and when this is coupled with the rap-idly increasing popularity of the 5-inch breech-loading all-steel rifle, our country notably may congratulate itself that ordnance is reverting to a plane which other nations mistakenly and at great cost abandoned, arid which the United States can readily attain.

Ships of the Minor Navies

Early in September of this year there sailed from England for the East five Chinese war-vessels of the latest types: the *Chih Yüan* and *Ching Yüan*, fast cruisers; the *King Yüan* and *Lai Yüan*, coast-defence ships; and a torpedo-boat as yet unnamed. Though the squadron was commanded by Admiral Lang, a captain in the Royal Navy temporarily serving under the Chinese government, the other officers were mainly, and the crews were wholly, natives who had passed through English cruising and training ships. The *Chih Yüan* was commanded by Captain Tang, who had under him nine English and fifteen Chinese officers and one hundred and fifty men; the *Ching Yüan* was in charge of Captain Yih, and eleven English and fourteen Chinese officers, with the same complement; while the other ships were officered and manned much the same way. There was, it is true, an English fleet surgeon, but each ship had its native medical officer and two chief engineers, one of whom was a Chinese. The *Herald* cable despatch stated:

> On leaving Spithead the fleet will proceed direct to Gibraltar, thence to Port Said, where it will take in coal; it will stop at Suez, Aden, Colombo (where it will coal again), Singapore, Hong-Kong, Chefoo, and Taku, joining at this place the fleet already assembled under Admiral Ting, and replacing there many of the foreigners by native officers. The voyage is expected to occupy seventy-two days—fifty-two at sea and twenty in harbour—and during this time the crews will be thoroughly practised in torpedo, gun, and other drills. This, of course, will involve a deal of hard work, such as would try the endurance of English sailors, but the Chinamen will be allowed a plentiful supply of beef and beer.

Modern cruisers and armed battleships requiring the highest intelligence to fight, torpedo-drills, beef and beer—and all for that outer barbarian whom our mobs murder just for a lark! Here is a lesson for Congressmen; here an example and a possible menace for this defenceless land.

The Chinese navy, though of recent growth, consists today, (as at time of first publication), of seven armoured and ten unarmoured ships of modern types, in addition to torpedo-boats, and to at least thirty other vessels which are not so obsolete as nine-tenths of the ships this country has in commission,

Nearly ten years ago the Chinese government realized that its wooden corvettes, gun-boats, and armed junks were no longer adapted to warfare, and ordered from the Vulcan Works at Stettin the two steel cruisers *Nan Shu* and *Nan Shen*. These are of 2200 tons displacement, and with 2400 horse-power have developed 15 knots speed; their armament is composed of two 8-inch and eight 4½-inch Armstrongs, and of lighter secondary pieces. In 1881 these ships were followed by the armoured battleships *Chen Yüan* and *Ting Yüan*, and by the steel cruiser *Tchi Yüan*. The battleships are built of steel, and have the following dimensions: length 296.5 feet, beam 59 feet, mean draught 20 feet, displacement 7430 tons. Their compound armour extends throughout a central citadel 138 feet long, and around a nearly elliptical redoubt situated at its forward end; the side armour is five feet wide, and has a thickness of 14 inches at the water-line, of 8 inches at the lower and of 10 inches at the upper edge; the protection to the redoubt is 12, and to the conning-tower 8, inches thick. The armament consists of four 12-inch Krupps, echeloned in pairs within the redoubt; of two 5.9-inch Krupps mounted forward and aft inside of machine-gun proof turrets; of eleven Hotchkiss revolving cannons, and a supply of Whitehead torpedo-tubes. The engines are of the three-cylinder compound type, and develop 7300 horse-power and 15.5 knots. The ships have double bottoms, minutely subdivided, and in addition to a cork belt forward and abaft the citadel a steel protective deck two inches thick curves to the extremities. The twin-screw steel cruiser *Tchi Yüan* is of 3200 tons displacement, and has two sets of two-cylinder horizontal compound engines, which develop 2800 horse-power and a speed of 15 knots; her dimensions are: length 236 feet 3 inches, beam 34 feet 5 inches, and draught 15 feet 9 inches. The entire underwater body is covered by a curved steel deck, which is 4 inches thick, and extends 4 feet 9 inches below the water-line; the space between this deck and the one above is used for coal-bunkers.

There are two machine-gun proof turrets on the fore and aft line, the base of the forward one being surmounted by a fixed tower armoured with 15-inch steel, which extends to a height sufficient to protect the base of the turret, its machinery, and gun-carriages. The armament is composed of two 8.27-inch (21 centimetre) Krupps in the forward turret, of one 5.9-inch (15 centimetre) Krupp in the after turret, of two similar guns on the main deck aft, of five Hotchkiss revolving cannons, and of a supply of Whitehead torpedoes, discharged through four above-water tubes. (Lieutenant Colwell, United States navy)

The swift protected cruisers *Chih Yüan* and *Ching Yüan* were built at Elswick; the unnamed torpedo-boat is of the *Yarrow* type; and the coast-defence vessels *King Yüan* and *Lai Yüan* were constructed at Stettin. The displacement of the cruisers is 2300 tons, length 268 feet, beam 38 feet, depth 21 feet, and draught 14 feet forward and 16 feet aft. Each vessel has two pairs of triple-expansion engines. Both the engine and boiler rooms are divided into water-tight compartments by transverse and longitudinal bulkheads, and the machinery is so arranged that either boiler can work on one engine or on both, and the change necessary to effect this can be made while the vessel is in motion. The result of this intercommunication between each engine and each boiler is that the vessel can proceed so long as any single boiler and engine are uninjured.

In the four trial trips, two with and two against the tide, with all their weights, armament, and Chinese crews on board, they attained an average speed of 18.536 knots.

The vessels are built of steel, and have two decks, the lower one consisting of four-inch steel plates, rising in the middle above the water-line and inclined at the sides so as to dip below it. The engines, magazines, rudder-head, and steering gear lie below, and are protected by this deck. The openings in the deck are encircled by coffer-dams, armoured with steel plates, inclined so as to deflect projectiles. The bows are formed and strengthened for ramming purposes. Additional protection is given to the vessel by a partition which is built on the protective deck parallel to the side of the ship; this encloses a space that is eight feet wide, and is subdivided into a great number of water-tight compartments for the stowage of four hundred and fifty tons of fuel. Both ships have double bottoms, minutely subdivided into water-tight compartments.

The armament consists of three 21-centimetre Krupp guns—two

mounted forward and one aft—all on centre-pivot, shield-protected Vavasseur carriages; of two 6-inch Armstrongs on sponsons, also Vavasseur mounted; of eight 6-pounder rapid-fire Hotchkiss; and of six Gatling guns. There are four above-water torpedo-tubes—two fixed (one in the bow, firing ahead, and one aft, pointed astern) and two training, one in each broadside.

There are two electric search-lights for each vessel, with a nominal power of 25,000 candles, while the cabins and rest of the ship are lighted with incandescent lamps. The *Army and Navy Gazette* writes:

> It is humiliating, but nevertheless an actual fact, that two of the cruisers of the Chinese squadron under command of Admiral Lang are superior in certain novelties of construction to any of our own vessels of this class. In point of speed the two unarmoured ships which have been turned out by the Elswick firm cannot be touched by our swiftest cruisers. They steam nearly nineteen knots an hour. The traversing and manipulation of their guns can be effected with such rapidity that when saluting the garrison at Portsmouth recently it appeared almost impossible that the guns could have been properly sponged between each discharge, the two bow guns especially keeping up a continuous roar. Only the two sponson 6-inch guns are from Armstrong's; they are mounted on Vavasseur carriages, and fitted with singularly simple breech apparatus. The other three heavy guns are Krupp's 21-centimetres (about 8¼-inch). These last are protected with a shield of entirely unique construction. It is of steel, and commencing from the trunnion ring spreads out into a wide shelter sufficient to accommodate the entire gun detachment. The sights are also under cover. The stern-chaser has a single shield; the two bow-chasers are included within one. The torpedo apparatus is most complete. In addition to the two tubes opening ahead and astern, which are well above the surface of the water, there are six others in connection with the torpedo-room.
>
> But the latest improvement which is observable on board is the steel armoured conning-tower, fitted with Lord Armstrong's patent telegraph and communications, for which a special royalty of four hundred pounds has to be paid. It is the most perfect scheme for conducting fighting operations that has ever come under our notice. A model for laying all the guns is prominently placed in front of the steering-wheel, which is under personal

command of the officer in charge. On the left are tubes and telegraphs by which he can converse with the officer in command of the gun detachment, and correct any mistakes observable in the laying of the guns. Then he can fire simultaneously, if desirable, or singly, if preferred. All stations on board are also in communication with this conning-tower. Hence the entire fighting power of the vessel, torpedoes and all included, is at the disposal of the officer in command within the conning-tower. Another useful modification has been effected in these vessels. The conning-tower, which is at the foot of the foremost fighting mast, has close to it the signal station, also protected with steel armour, so that the signaller therein is absolutely secure, and close to the commanding officer, from whom he receives and to whom he communicates outside signals.

The torpedo-boat built by Yarrow is said to be the fastest of its size that has ever been launched, as it has reached a speed of about twenty-eight knots an hour. It is armed with two fixed 14-inch torpedo-tubes in the bows, and one 14-inch training-tube on deck abaft the funnel. It is also supplied with a powerful armament of Hotchkiss and Gatling guns, and a strong electric search-light so arranged as to be worked either from the conning-tower or from the deck.

The *King Yüan* and *Lai Yüan*, built by the Vulcan Company at Stettin, are powerful vessels, effective for either coast defence or distant sea service. Their principal dimensions are: length 269 feet, beam 39 feet 4 inches, depth 25 feet 6 inches, mean draught 16 feet 8 inches, and displacement 2900 tons. They are built entirely of steel, with double bottoms extending two-thirds of the length, and the underwater body is divided by bulkheads into sixty-six water-tight compartments. The armour protection is compound, and consists of a belt six feet wide extending the length of the machinery and boiler space, having a maxi-mum thickness of 9.5 inches at and above the water-line, and a minimum thickness of 5.1 inches. This belt is terminated at either end by athwartship armoured bulkheads, 5.1 inches thick. At the forward end of the belt is a circular revolving turret eight inches thick, on top of which is the conning-tower, with an armour protection of six inches. The under-water body is protected by a complete steel protective deck, 1.5 inches thick over the top of, and three inches thick forward and abaft, the belt. A partial cork belt above the protective deck gives additional stability. The engines consist of two sets of three-cylinder compound type, situated in two separate compartments, driving twin screws, and developing 3400 horse-power with forced draft. The boilers, four in number, are placed in two separate compartments. A speed of about sixteen knots was attained. The armament consists of two 21-centimetre (8.27-inch) Krupps mounted in the turret; of two 15-centimetre (5.91-inch) similar guns carried in recessed ports; of two 47-millimetre Hotchkiss rapid-fire guns; of five 37-millimetre revolving cannons; and of four torpedo-discharging tubes—three above-water and one in the bow below the water-line.

As additions to the lightly armoured gun-boat *Tiong Sing*, built in 1875, China ordered this year from the Vulcan Company two heavy coast-defence vessels of 7000 tons displacement and 6000 horse-power, and laid down at Foochow an armoured gun-vessel. The *Tshao Yong* and *Yang Wai* are steel cruisers built at Elswick, of 1350 tons displacement and 2400 horse-power; they have developed sixteen knots, and are armed with two 10-inch and four 4½-inch Armstrongs, with a secondary battery of two lighter pieces and six machine guns. The *Fee-chen*, a small steel cruiser built at Sunderland, England, has triple-expansion engines, and is expected to develop thirteen knots. Her armament consists of two 6-inch Armstrongs and four lighter guns; she is also fitted to do cable work. Three cruisers of the *Nan Shu* type are being constructed in Chinese dockyards, besides several of the *Kuang Chen* class of gun-boats.

The Japanese navy consists of forty vessels, of which eight only are modern. The classified armoured fleet includes five ships, among them the *Adsama Kan*, formerly known as the *Stonewall Jackson*; none of these is of any importance except the central battery ship *Fu Soo*, which was launched in 1877. In January of this year, however, the Japanese government ordered from the Société des Forges et Chantiers de la Méditerranée two coast-defence vessels, to be built on the plans of M. Bertin, constructing engineer of the Japanese navy. They are to be built entirely of steel, on the cellular plan, with two longitudinal and twelve transverse bulkheads. Their principal dimensions are: length 295 feet 2 inches, beam 50 feet 6 inches, depth 34 feet 9 inches, draught aft 21 feet 2 inches, displacement 4140 tons. The armament proposed is one 12.6-inch (32-centimetre) breech-loader, eleven 4.72-inch (12-centimetre) breech-loaders, six rapid-fire guns, twelve revolving cannons, and four torpedo-tubes—one in the bow, one in the stern, and one each broad-side. Two independent triple-expansion engines, driving twin screws, and required to develop 5400 indicated horse-power with forced draft, and 3400 with natural draft, supplied with steam by six three-furnace boilers in two groups, furnish the motive power.

The estimated maximum speed is sixteen knots. A heavy protective steel deck and a complete surrounding arrangement of coal-bunkers protect the engine and boiler space and magazines. The complement of officers and men will number four hundred. In March, 1887, a small armoured gun-vessel, designed by the same official, was laid down at the Ishikawa-Shima dockyard, Japan. The displacement is 750 tons, length 150 feet, and beam 25 feet. (*Recent Naval Progress.*)

Of the unarmoured vessels, the sister-ships *Naniwa Kan* (already described) and *Takatschio* are at present the most important, though six modern cruisers now under construction in Japanese dockyards will soon be added to the fleet. The navy is manned and officered exclusively by natives, and the service is well administered and popular. Owing to possible complications with China, coast defence has become a live national question, and the wealthy Japanese are subscribing large sums for ships and forts. In addition to these voluntary contributions, the new tax which has been imposed will enable Japan to put herself in an excellent condition for attack or defence.

The other navies not described in these pages have afloat or under construction but few modern ships-of-war. Still, there are vessels in the minor services which ought to be briefly described. One of these, the *Almirante Brown*, of the Argentine navy, is a twin-screw, central-battery steel ship which was launched in 1880. Her dimensions are: length 240 feet, beam 50 feet, draught 20 feet 6 inches, and displacement 4200 tons. With 4500 horse-power she attained 13.75 knots, and her coal endurance is given as 4300 knots at 10 knots speed. Her armament is made up of eight 8-inch and six 4½-inch Armstrongs, and of four machine guns; the armour is compound, nine inches thick on the belt and eight inches on the battery. There is also building in England for this government a central casemate steel cruiser of 4400 tons displacement. The armour on the casemate is to be compound, ten inches thick, and the armament is to be composed of eight 8-inch breech-loading Armstrongs, with a secondary battery of rapid-fire guns and torpedo-tubes. The estimated speed is fourteen knots. In addition to these two vessels the Argentine navy has two small coast-defence turret-ships, one 14-knot steel cruiser (the *Patagonia*, which is similar in appearance to the United States steamer *Atlanta*), six gun-boats, eleven torpedo-boats, and a few other vessels of an unimportant character.

The Brazilian navy has, exclusive of her capital torpedo-boat flotilla, over fifty vessels, of which a dozen are classed as armoured. These last are mainly medium draught, coast-service turret-ships and river monitors, though among them are the *Riachuelo* and *Aquidaban*, twin-screw armoured cruisers, and the *Solimoes*, an armoured battleship. The *Riachuelo* made a sensation when she first appeared, and is still one of the most formidable vessels in the world. She is built of steel, and has the following dimensions: length 305 feet, beam 52 feet, draught 19 feet 6 inches, displacement 5700 tons. Her armour is compound, eleven inches thick on the belt and ten inches on the turret, conning-tower, and redoubt. She has also a steel deck, which curves forward to strengthen the ram, and aft to protect the steering gear. Her armament consists of four 9-inch 20-ton Whitworths (Armstrong altered) mounted in two echeloned turrets, and of six 5½-inch guns carried under cover in the superstructure. Her secondary battery includes fifteen machine Nordenfeldts and five above-water torpedo-tubes. With 7300 horse-power she attained a speed of 16.71 knots, and is credited with a coal endurance of 4500 miles at 15 knots speed.

The *Aquidaban* is of the same type and general appearance as the *Riachuelo*, but of smaller dimensions. Her length is 280 feet, beam 52 feet, draught 18 feet, displacement 4950 tons. The compound armour is from seven to eleven inches in thickness, and seven feet in width on the water-line belt, and is ten inches thick on the conning-tower and on the oval redoubts which protect the bases of the two echeloned turrets. The armoured deck and redoubt roofs are built of steel, from two to three inches thick. The armament consists of four 9-inch 20-ton guns mounted in the turrets, and of four 70-pounders carried under the superstructure. The secondary battery is made up of fifteen 1-inch Nordenfeldts and five above-water torpedo-tubes. She developed on trial 6251 horse-power and a speed of 15.81 knots, and made on the voyage from Lisbon to Bahia 3600 knots in 13 days and 17 hours, and from Bahia to Rio Janeiro 750 knots in 2 days and 20 hours. The average speed for the passage from England to Rio was nearly eleven knots on a daily coal consumption of forty-three tons.

The Chilean navy has the two iron-armoured, twin-screw, central-battery ships Almirante *Cochrane* and *Blanco Encalada*, and the lightly armoured turret-ship *Huascar*. The *Almirante Cochrane* and *Blanco Encalada* are 210 feet in length, 45 feet 9 inches in beam, 19 feet 8 inches in draught, and 3500 tons in displacement. The former carries four 9-inch and two 7-inch breech-loading Armstrong rifles, four lighter pieces, and seven machine guns. Before the alterations and repairs lately made, the *Blanco Encalada* had six 12-ton muzzle-loading Armstrong rifles, four lighter pieces, and seven machine guns. The *Huascar* was built in 1865, and is a slightly protected iron ship of 2032 tons displacement, 1050 horse-power, and 12 knots speed. Her battery consists of two 10-inch muzzle-loading Armstrongs and two 40-pounders. Her wonderful record on the west coast of South America has made her name as familiar in the mouth as a household word, and whatever may have been the justice of the war, there never can or will be a question of the superb courage with which she was fought by her gallant officers and crew. Chilli has three wooden corvettes, the *Chacabuco*, *O'Higgins*, and *Pilcomayo*, one composite corvette, the *Magellanes*, one steel cruiser, the *Esmeralda*, five gun-boats, two paddle steamers, one despatch-boat, one transport, and eleven torpedo-boats. In April, 1885, The *Esmeralda* ran from Valparaiso to Callao, 1292 miles, in one hundred and eight hours, the engines during the last eight hours barely turning over. In the exhaustive trials made before her departure from England the highest speed attained was 18¼ knots per hour. The *Esmeralda* is said to be at present, (as at time of first publication), in an inefficient condition, both as regards her speed and battery power. In November, 1886, the Chilean government gave the Armstrong firm an order for a powerful, partially-protected steel cruiser, which is to be of 4.500 tons displacement, and to develop 19 knots speed. Her armament is to consist of two 10-inch, one 8-inch, and two 6-inch Armstrong breech-loaders, with a secondary battery of four 6-pounder rapid-fire guns, eight Hotchkiss revolving cannons, and eight torpedo-tubes.

Appendix 1
Submarine Warfare

The practicability of submarine navigation was established by the Dutch over two hundred and fifty years ago. Then, as now, its underlying idea, its claim for recognition, was the advantages the system gave in marine warfare. Nor is its battle value overestimated; for such a boat, if successful, exercises an influence that is great in material uses, that is enormous in moral effects. Its development has been slow; for though the problem was solved long ago, no practical results were attained until within the last thirty years. During the late war submarine boats were for the first time employed with such sufficient success that the great maritime powers have considered the type to have an importance which justified investigation. They reached this conclusion because no plan of defence exists which could defy the operations of a weapon that attacks not only matter but mind.

There is no danger which sailors will not face; because their environments are always perilous, and their traditions are rich with glorious records of seeming impossibilities overcome by pluck and dash. They are willing always, even against the heaviest odds, to accept any fighting chance. They know that the unexpected is sure to happen. The spirit that made Farragut take the lead of his disorganized line in Mobile Bay still lives; his clarion call of "*Damn the torpedoes! Follow me!*" is a sea instinct, born of brine and gale, which never dies.

Whatever coast fighting or port blockading may demand, sea battles are unchanged. History teaches that ships always closed for action, and that vessels fighting each other from beyond the circling horizons, or hull down, with long-range guns, are the dreams of shore inventors. Guns and ships have changed, but men and the sea are changeless. The fighting distance of today is not much greater than it was in Nelson's or in Perry's time; and the next naval war will surely prove that battle will be nearly as close as in Benbow's age, when the gallant tars combed innocuous four-pound shots out of their pigtails, and battered each other within biscuit-throwing distance with deftly shied chocking quoins.

It is fortunate, in the interest of good, square fighting, that the operative sphere of submarine boats is limited to coast work. Fortunate, because while the bravery and the grit are the same, the threatening of a danger which cannot be squarely met is apt to benumb the heart of the stoutest. A sailor hates to run; he does not care to fight another day when the chance of the present is open before him; but of what avail are the highest courage and skill against a dull, venomous dog of an enemy who crawls in the darkness out of the deeps, and, silently attaching a mine or torpedo, leaves his impotent foe to sure destruction?

Submarine mines may be countermined; when necessary, defied; guns may be silenced and torpedo-boats so riddled by rapid-fire guns that they will be disabled beyond the radius of their effective action; automatic torpedoes may be checked by netting, or by the prompt manoeuvrings of the attacked vessel; ship may always fight ship. But what is the chance for brain 'or brawn against a successful submarine boat, when the mere suspicion of its presence is enough in itself to break down the blithest, bravest heart of oak. It is here that their moral effects are enormous.

The history of their development may be briefly told. In 1624 Cornelius Van Drebble, a Hollander, made some curious experiments under the Thames. His diving-boat was propelled by twelve pairs of oars and carried a dozen per-sons, among them King James I. In 1771 Bushnell, of Connecticut, constructed a boat which Washington described in a letter to Jefferson as being:

> A machine so contrived as to carry the inventor under water at any depth he chose, and for a considerable time and distance with an appendage charged with powder, which he could fasten to a ship, and give fire to it in time sufficient for his returning, and by means thereof destroy it.

Fulton borrowed Bushnell's idea, and in 1801 experimented successfully with it in the Seine. He descended under water, remained for twenty minutes, and after having gone a considerable distance, emerged. In 1851 a shoemaker named Phillips launched in Lake Michigan a cigar-shaped boat forty feet long and four feet in its greatest diameter. This was his first attempt, but in the course of a few years he so far perfected his arrangements for purifying the air that on one occasion he took his wife and children, and spent a whole day in exploring the bottom of the lake. In the history of these boats, as told in the report of the Board on Fortifications, Phillips afterwards descended in Lake Erie, near Buffalo, and never reappeared.

Many other attempts were made, the most successful being that of a Russian mechanic, who in 1855 built a diving-boat which was under such perfect control that he could remain submerged for eight hours. The boat which sank the *Housatonic* was a remarkable submarine vessel-; it was about thirty-five feet long, built of boiler iron, and had a crew of nine men, of whom eight worked the propeller by hand, while the ninth steered and governed the boat. She could be submerged to any desired depth or could be propelled on the surface. After various mishaps she went out of Charleston harbour, attacked and sank the United States steamer *Housatonic*, then on blockade duty; as she never returned, it is supposed that the reflex action of the torpedo destroyed her.

In the report quoted above the results already attained in submarine navigation are thus summarized by Captain Maguire, U.S.A.:

1. Submarine boats have been built in which several persons have descended (with safety) for a great distance below the surface of the water.

2. Submarine boats have been propelled on and under the surface in all directions.

3. The problem of supplying the necessary amount of respirable air for a crew of several persons for a number of hours has been solved.

4. Steam, compressed air, and electricity have been used as the motive power.

5. The incandescent electric light has been used for illuminating the interior of submarine boats.

6. Seeing apparatus have been made by which the pilot, while under water, may scan the horizon in all directions.

7. A vessel has been in time of war destroyed by a submarine boat. The latter, it is true, was also sunk, but it was for reasons that are no longer in force.

As yet no perfectly successful boat of this type has been tried in any naval war, but there is no question that they will be used at the very first opportunity. Compared with a surface boat, the submarine has the following advantages:

1. It does not need so much speed. The surface boat demands this quality so as to get quickly within striking range of its torpedo, and then to escape speedily out of range of machine guns, etc.

2. Its submersion in the presence of the enemy prevents the engines being heard.

3. There is no smoke nor glare from the fires to cause its detection.

4. The boat and crew, being under water, are protected from the fire of machine guns and rifles.

5. It is enabled to approach the enemy near enough to make effective even an uncontrollable fish torpedo.

6. It can be used with safety as a reconnoitring or despatch boat.

7. It can examine the faults in the lines of submarine mines, and replace mines exploded in action. Abroad, the Nordenfeldt boat has awakened the most interest, and here the American submarine monitor holds the first place.

The form of the Nordenfeldt boat is that of a cigar or of an elongated cylinder tapering away to a fine point at each end. The outer case, built of stout steel, is calculated in its construction to resist such a pressure as would enable the boat to descend even beyond a depth of fifty feet, although that is set as the maximum for its diving operations. The cigar shape does not at first sight com-mend itself, even in the eyes of nautical men, on account of its supposed tendency towards a rolling motion. The experience, however, gained with the boat exhibited for the benefit of naval experts at Carlscrona, in September, 1885, has shown that very good sea-going qualities can be developed in a craft built upon, such lines; for this small vessel has weathered more than one gale in the Baltic, to say nothing of the severe storm it encountered at the entrance to the Kattegat when proceeding from Gottenburg to Copenhagen for the experimental trials.

This quality results from the fact that each end of the boat forms a tank, which is filled with water, and as there is no extra buoyancy in those directions, and consequently no tendency to lift at those parts as with an ordinary vessel in a sea-way, the vessel rises and falls bodily instead of pitching. It has been found that by going at a moderate

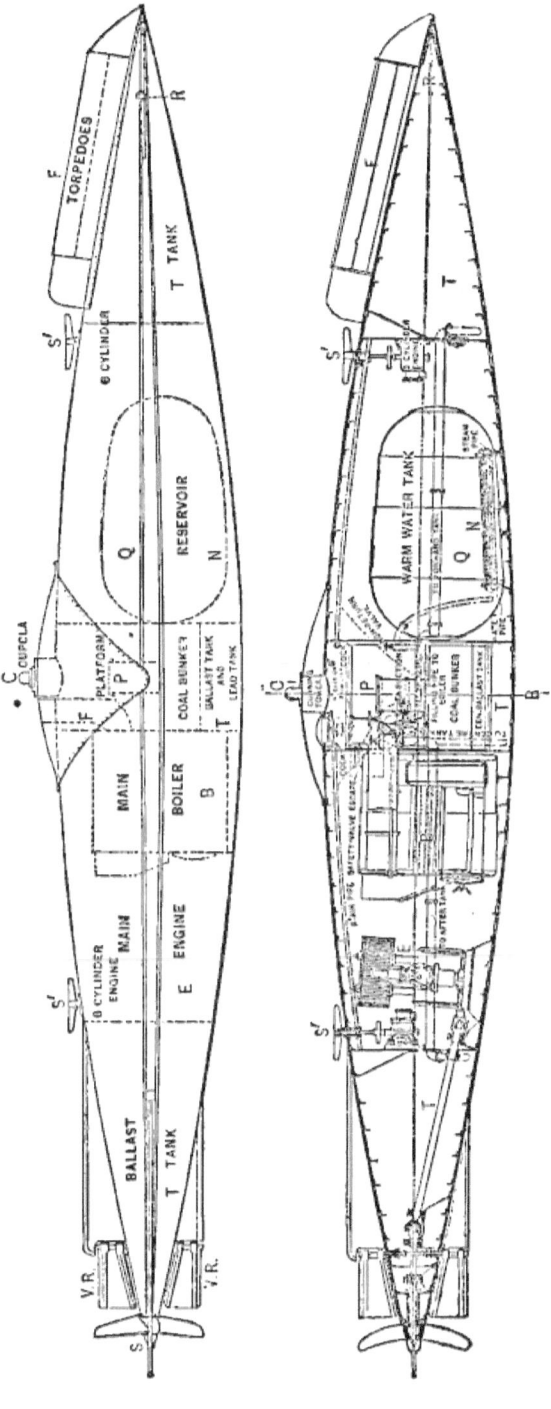

Longitudinal plans of Nordenfeldt boat

speed and taking the seas a point or so on the bows the boat makes very good weather, as the waves, breaking on the snout, sweep over the fore part and expend their force before any portion of them can reach the central section.

Steam, which is employed as motive power, is perfectly trustworthy as an agent. There is nothing about its action, or the appliances connected with it, that is beyond the grasp of an ordinary engineer, whereas such can hardly be said as yet in respect either to electricity or the other agencies by which inventors have sought to obtain motion. The difficulty, however, has always been how to retain steam pressure for any great length of time without carrying on combustion. This in the Nordenfeldt boat is secured in the following ingenious manner: A large reservoir or hot-water cistern (marked Q in the plate) is placed in the fore part of the boat, in communication with the boiler. The steam from the latter passes through a number of tubes in the reservoir N, thus raising the temperature of its contents until the pressure stands at the same degree in both. While the boat is at the surface, the maximum pressure once attained, as long as combustion is carried on, supplies quite enough steam both for driving the engines at full speed and for maintaining the contents of the cistern in the proper superheated condition. When the boat is submerged and the furnace doors are closed combustion ceases, and the steam given off by the hot water in the boiler and cistern is sufficient to keep the engines going for several hours. Submersion to the various depths required is secured by the motion of the vertically acting screws, S S, driven by small three-cylinder engines. The boat is so ballasted as always to have spare buoyancy, and while a few revolutions of the screws will send her under water, the arrest of their motion is all that is required to bring her to the surface again. In this arrangement, as even the non-technical reader will readily understand, there is a great element of safety, the rising motion being entirely independent of any machinery which might refuse to act at the required moment. Another advantage is also gained in the ease with which the horizontal position is maintained by regulating the speed of the screws. To assist in keeping this position there is a horizontal rudder or fin, R, at the bows, which, by a very ingenious arrangement of a plumb weight with other mechanism in connection with the steering tower, works both automatically and by hand. The torpedoes are carried on the outside of the boat, as shown at F. They are Swartzkoph or Whitehead, as the case may be, and are released by electrical action under the control of the captain, standing on the platform at P. C is a cupola of stout glass by which a view is obtained occasionally when the boat is running submerged.

Construction Details.—The following are the dimensions of the Turkish boat: length 100 feet, beam 12 feet, displacement 150 tons, speed 12 knots, and coal endurance sufficient for travelling 900 miles. The engines (E) are of the ordinary inverted compound surface condensing type, with two cylinders, and with 100 pound pressure indicate 250 horse-power. The circulating and air pumps being actuated by a separate cylinder, the main engine is left free to work or not, while a vacuum is always maintained to assist the various other engines with which the boat is fitted. In this respect it should be mentioned that all the engines are specially designed with such valve arrangements as will make the utmost use of the vacuum, it having been found that while the boat is running beneath the surface as much power can be developed below the atmospheric line as above it.

The boiler, B, is of the ordinary marine return-tube type, with two furnaces, and the heating surface is about seven hundred and fifty square feet. The tanks at each end of the boat contain about fifteen tons each, and there is a third of seven tons capacity at the bottom of the central compartment for regulating buoyancy. The coal is stored around the hot-water cistern as well as at the sides of the boiler and over the central ballast tank.

Three men and the captain can efficiently work this boat, although she may carry a crew of seven, who could remain in her for over seven hours beneath the water without experiencing any difficulty in respiration. No attempt is made as in some systems to purify the atmosphere by chemical means, as it is said to be quite unnecessary.

The Practical Management.—The boat is operated in the following manner: Steam having been raised to the required pressure, the funnel is lowered, and water is let into the ballast tanks to bring the craft down to the proper trim for action. In this condition the screws, S S, are sufficiently under water to obtain the requisite thrust. The boat may still proceed at the surface for some time if the enemy be distant, but the conning-tower should be closed, and the cupola hatch and the furnace doors shut, before there is any chance of discovery. The vertically acting screws being started, the boat is then submerged to the cupola, and continues approaching until, according to circumstances, it becomes prudent to disappear entirely. The direction is taken at the last moment, and maintained by compass until within striking distance, when a torpedo is released, and the boat immediately turns in another direction.

In May of this year, (as at time of first publication), there was launched at Barrow a Nordenfeldt boat 110 feet in length and 13 feet in diameter. The engines are capable of developing good power, and a speed of 12 knots on the surface was realized. The boat was tried on the Bosporus during July under government supervision, and as these were satisfactory, it seems likely that a number of similar vessels will be built next year for the Ottoman navy.

The original submarine monitor *Peacemaker* is well known through its trials on the Hudson River in 1886, but since then so many improvements have been made in the direction of increased efficiency that it is confidently expected the boat just designed will surpass its former successes. It must be understood in the beginning that its essential principle remains the same, all the important improvements being the outgrowth of the experience gained in previous experiments.

Broadly defined, the new craft has a midship section, which through its high centre of buoyancy and low centre of gravity gives great stability of form, or, to make it plain to the non-technical reader, it differs from the ordinary cigar and tortoise shaped boat in being more nearly like the section of a pear, the apex of which forms the keel. Its longitudinal section is not unlike the form generally used, though the lines are such as have been found to give the form of least resistance and the highest speed.

It is built of steel, with frames and spacings sufficient to stand the pressure of the lowest depth to which the boat is or can be expected to go. The old dimensions were: length 30 feet, depth 7 feet, and beam 8 feet. In order to obtain increased speed the present vessel will be 50 feet in length, 8 feet in beam, and 8 feet in depth, with a displacement of from thirty-five to forty tons, or an amount sufficient to carry the weights of the interchangeable boiler, of the sixty horse-power engine, and of the provisions and fuel necessary for a surface cruise of one week, and, when necessary, for a constantly submerged cruise of twelve hours.

The advantages claimed for the new boat are that she is so self-sustaining as not to need the assistance of any other vessel; that she is not an accessory, but has in herself all essentials of defence; and that she answers all possible necessities for submarine work of any kind whatever, whether in peace or war. The increased speed will, it is hoped, give her power to attack modern vessels under way. When submerged, as was proved last summer, she sent no bubbles of air to the surface, and had neither a wake nor a wash to militate against the possibilities of an absolutely secret attack. Besides these advantages, the boat is

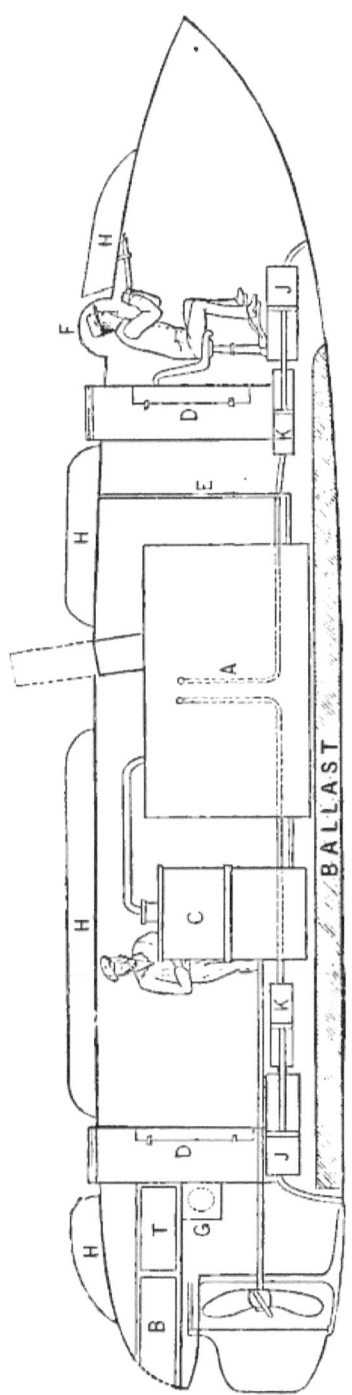

Sketch submarine monitor Peacemaker

said to be a safe surface-cruising vessel, forming no target for the destructive action of an enemy's attack, and at the same time having a capacity for disappearing so readily under water and avoiding the possibility of discovery that the enemy will be unable to tell when, where, or how the assault upon him may be made.

As in a former trial an accident proved the danger of an exposed conning-tower, the Submarine Monitor Company have provided a fin or guard for protecting the new helmsman's lookout and companion-hatches. The waterlock appliance employed in the original boat has now an additional use in supplying a mode of egress and ingress, the opening being made telescopic, so as to permit surface runs in comparatively rough water. When submerged, the smoke-stack acts telescopically, and is closed with a water-tight valve. To avoid the necessity of divers going out of the boat when under water, there are various openings at places in the exterior skin to which rubber sleeves or arms, with a radius sufficient to cover almost all practical necessities, will be fitted. These apertures do not constitute planes of weakness or danger, because they are normally closed by stout water-tight dead-lights.

The Westinghouse engine is employed, as its construction prevents, by the packing used, any radiation of heat and the consequent elevation of temperation below. The air-tight doors and bulk-heads work laterally, and the conning-dome is made of steel, with such apertures as will enable the helmsman to have, when on the surface, an all-round view, and when submerged, a sufficient light to let him in the daytime read, at a depth of thirty feet, the time by his watch.

Should the necessity arise, when submerged, the purity of the atmosphere below is preserved by passing the air through caustic soda, thus eliminating carbonic acid gas, and by reinforcing the loss of oxygen from tanks of compressed air. In the original experiments the boat was frequently submerged six hours at a time, and the crew of two men had no other air supplied than that which the boat carried down with her.

Besides these chemical means there are rubber tubes floated by buoys, with nozzles which protrude above the wash of the surface water. There is in each tube an automatic valve, which prevents water coming through the pipe at the time the air is being pumped in, and the depth below the surface to which outside air can be supplied is limited only by the length of the pipe.

In the plate, A represents a patented inter-changeable boiler, in which either hydro-carbon-ate fuel or caustic soda can be used, in both cases steam being the motive power. The interior boiler for the use of the caustic soda is surrounded by a jacket, into which the steam exhausted from the engine can be used before it becomes so saturated as to create a back pressure on the engine, that is, for a period of twelve hours, when this limit is attained, and the surface is reached, the soda can be blown off into an outer receptacle provided for the purpose, and then reheated and recharged. The hydro-carbon fuel is ordinary mineral oil, carried in tanks of sufficient capacity for a surface run of a week. It may be emphasized as an important fact that this method of exhausting into the jacket of the boiler avoids the possibility of any bubbles appearing on the surface, as was notably the case with the earlier Lay boats.

Before diving, the caustic soda, which has been already heated by the combustion of the oil to the proper degree, acts in place of the ordinary fuel, thus constituting a sort of perpetual motion, until the point of saturation is reached, and back pressure in the engine results.

The boat, when on the surface, is run with the oil fuel, but as soon as it becomes necessary to dive this fire is extinguished, the after-hatch is opened by unlocking the door of the bulkhead separating the after from the bulkheaded end of the vessel, and by a system of fans the hot air from the fire-room is driven outboard. Then the after telescopic hatch is reefed and secured, the soda is thrown from the receptacle where it has been heated into the jacket of the caustic-soda boiler, the fires are put out, the smoke-stack is taken in and securely fastened, and the machinist, leaving the engine-room, goes through the bulkhead door into the forward compartment, where he has complete control of the machinery and boiler by means of a duplicate set of gauges and levers. In case of an attack, the man detailed for operating the main torpedo is left in the after compartment, where he has access to that weapon and to the buoy, reel, and other mechanical appliances employed in its operation.

The helmsman, who controls the steering apparatus that governs the horizontal and perpendicular rudders, also operates with his feet the levers which are connected by links to the throttle that supplies steam to cylinders K K. These last function like the Westinghouse brake, and are connected with pistons to the cylinders J J. Through their agency water is at will admitted into or forced out of the larger receptacles, either from one end or from both ends simultaneously. The effect of discharging water is of course to increase the buoyancy of the vessel; and of admitting it, to decrease this quality so that without changing structural weights the boat is enabled to rise or sink perpendicularly, or, by admitting more water in one end than in the other, to take a downward or an upward course. Though this does away with the necessity of the horizontal rudder, it is kept as an additional resource for steering. In case of accident to the connecting pipes or machinery the vessel is supplied with water receptacles and hand-pumps, which are able to govern its submergence so that should all other mechanism break down the boat is so completely under the control of the operator that it can at all times be brought to the surface. As an additional safeguard, there is on the outside of the boat a quantity of ballast which can be readily detached by the arms or sleeves previously described, and so effectively that the reserve buoyancy thus gained will alone carry the boat to the surface.

In addition to the main torpedo and buoy resting in the cylindrical apertures aft, other torpedoes, connected by spans, are carried on deck. The method of their employment in attack is to go under the body of the vessel athwartships, and to liberate them. As they are fitted with magnets, they will, it is claimed, when freed, attach themselves to the bilges of the enemy's vessel, while the *Peacemaker* can continue her cruise and let them act automatically, or, backing off to a distance greater than the depth of water in which she then is, safely explode them by conventional electrical appliances. With the increased speed of the present boat there are various methods of attacking vessels of war when under way, among them one which is somewhat similar to that described above.

The *Peacemaker*, when under the body of the vessel athwartships, would liberate a buoy, B, that is connected with a torpedo, T, by a chain, the length of which depends upon the depth beneath the buoy the torpedo is desired to float. The steel tow-line to the torpedo is payed out from reel G to a sufficient length, and then by going ahead with the boat the torpedo is drawn close under the opposite side of vessel from buoy B. In this position the torpedo can be exploded by electricity.

If necessary, by liberating buoy B, while crossing the bow on the starboard side of the fore-foot of a vessel, the forward motion will draw the torpedo, T, close in to the opposite side; then, by a system of push-pins on the torpedo, the operator learns that it is in close contact and ready for explosion by electricity. Should the enemy's vessel be at anchor the tide can be employed for the purpose of bringing the buoy on one side of the vessel while the torpedo is on the other.

The boat is supplied with the ordinary incandescent lights, or apparatus for lighting the interior for night attacks.

Torpedoes

America has contributed to modern warfare many of its most valuable inventions. In the decade of 1850-60 the steam frigates of the *Merrimac* class revolutionized the naval constructions of the world, and became the models for the warships of the great maritime powers. In the same period our coast defences reached the high-water mark of modern development, and, soon to be crystallized, there were seething in the brains of American inventors ideas of guns, ships, and projectiles which made history. Though today our created contributions to quick peace through arrested or irresistible war are meagre, still many of the theories which make possible modern ordnance and ships are the fruits of American genius and industry.

Is the future to be as fertile in thought and deed? Are the destroyers of Ericsson, the dynamite safety shells of Hayes, the guns of Zalinski, the torpedoes of Howell, Sims, or Berdan, the turrets of Timby, the submarine monitors of Tuck, the gun-carriages of King or Buffington, the ordnance of Sicard, Benét—are these to prove that Yankee brain and brawn are potent yet for the mastery of the problem?

The country has no plainer duty than to foster by every care American ideas working in national ways of thought. It is rich, public sentiment is ripe and responsive, and Congress should encourage in peace the experiments which may make war impossible. In the question of ship armament and sea-coast fortifications notably, the value of torpedoes is now so generally recognized that the definite selection of some type has attained an importance which demands most careful consideration. All experts agree that they are vital, but there is not that consensus of opinion which within limits affirms exactly what should be done.

The fortification board in their report say:

It is not generally considered possible to bar the progress of an armoured fleet by the mere fire of a battery; some obstructions sufficient to arrest the ships within effective range of the guns is necessary. The kind of obstruction now relied upon is the torpedo, in the form of a submarine mine, and, except in special cases, exploded by electric currents which are so managed that the operator on shore can either ignite the mine under the ship's bottom, or allow the ship to explode it by contact. In deep channels the submarine mines are buoyant; in comparatively shallow waters they are placed upon the bottom—the object in both cases being to touch or nearly approach the hull of the vessel. Submarine mines are not accessories to defence, but are essential features wherever they can be applied.

The senate committee on ordnance and warships reported:

Concerning another class of torpedoes, 'fixed' or 'anchored' or 'planted,' technically known as submarine mines, there is a great popular misapprehension. Their value is greatly overestimated. They require picked and trained men for their management, electrical apparatus for their discharge and for lighting up the approaches, stations on shore secure against sudden assault, a flanking fire of canister and case shot and of machine guns (themselves protected), light draught picket-boats, and the overshadowing protection of armoured forts and heavy guns. None of these things can be extemporized. The submarine mine alone is of little use, and it must accompany, not precede, more costly and less easily prepared means of defence.

There is, however, a more definite agreement as to the value of torpedo-boats. The fortification board declare:

Among the most important means of conducting an active defence of the coast is the torpedo-boat, which, although recently developed, has received the sanction of the nations of Europe, each one of which now possesses a large number of these vessels. Their use will be quite general. First, in disturbing blockades, and preventing these from being made close, as no fleet would like to lie overnight within striking distance of a station of these boats; secondly, in attacking an enemy's ship enveloped in fog or smoke; thirdly, in relieving a vessel pursued by the enemy; and fourthly, in defending the mines by night and by day against attempts at countermining, and in many other ways not necessary to recapitulate.

Impressed with the utility of this mode of defence, the Board recommended the construction of one hundred and fifty of these boats, and the organization of a special corps of officers and men from the navy trained to their use.

In England, Commander Gallwey does not hesitate to say that the torpedo-boat is for harbour defence so superior to the submarine mine that he would not be surprised if before long it superseded the latter altogether. In France, Charmes insists that an armoured vessel will run the most serious risk if a torpedo-boat is allowed to approach unobserved to within one thousand to fifteen hundred feet; that the torpedo will surely triumph over the ironclad, and that armour has been vanquished, not by the gun, but by the torpedo.

A Naval Reserve

Among the problems to be solved by an efficient naval administration there is none more difficult or of greater importance than the formation of reserves of seamen. Our late war exposed the nation's weakness in sailors. At the beginning of hostilities the fleet, on paper, consisted of forty-two ships of all classes, mainly sailing-vessels, with a few paddle-wheel steamers, and less than ten screw-vessels with auxiliary power. Its *personnel* comprised seven thousand of all grades. And yet, to blockade a coast of over three thousand miles in length, the secretary of the navy had at his disposal but three effective vessels, and a reserve of only two hundred seamen on all the receiving-ships and at all the naval stations.

As late as the first of July, 1863, there were not men enough to carry out efficiently the work imposed upon the navy, and of the thirty-four thousand blue-jackets twenty-five thousand were landsmen. Secretary Welles, at the end of the same year, complained that there were no reserve seamen, that the supply for immediate and imperative duties was so inadequate that one of the largest and fastest steamers destined for important foreign service had been detained for months in consequence of the need of a crew, and that many other vessels were very much short of their complements. The cause of this was want of foresight, of prudence, of national commonsense even. We did not lack the material from which crews could have been drawn, for in 1860 over seventy-five thousand men sailed in the American merchant marine, fifty thousand of whom, under any system of enrolment suited to our national instincts and prejudices, would, before the end of 1861, have been available for duty on shipboard.

In peace there had been no organization, so when war came we were almost helpless, and as late as the end of 1863 not twenty per cent, of the men who should have been ready for service were in government ships. Let *doctrinaires* theorize as they may, this was not the fault of our maritime class, for thousands of sailors and fishermen who had already entered the army were by force of law denied the opportunity either of enlisting in, or of being transferred to, the navy. In addition, the operation of the draft was made detrimental to the naval interests of the country, for it violated the Act of May, 1792, which exempts from military duty all mariners actually employed in the sea service of any citizen or merchant within the United States. Furthermore, the government unjustly discriminated against the seaboard towns, for not only was the seafaring class, which is fostered and cherished by all maritime governments, withdrawn from the element to which it has been accustomed, but in addition sailors actually afloat were taken from their ships and compelled, under the penalty of law, to enter the land service. It was not until 1864 that Congress finally enacted the law which enabled seamen serving as soldiers to be drafted into the navy.

How different would have been the state of affairs had there existed in 1861 some system of government administration as to the creation of naval reserves, or, more far-reaching still, had we been free from that illogical distrust which possessed the whole country! The fear of too much centralization was the stock in trade of professional patriots, and the people, hampered by traditions which had come down to us from our English ancestors, saw in any attempt towards efficient war preparation in times of peace all the dangers they had been taught to believe existed in standing armies.

England acted more wisely, for she had been taught a grim lesson by her adversities, and without fear we might have profited by her example. In the history of the Peninsula war, Napier, after picturing the horrors of the fearful April night when Badajoz was stormed, asked, bitterly:

"And why was all this striving in blood against insurmountable difficulties? Why were men sent thus to slaughter when the application of a just science would have rendered the operations comparatively easy?

"Because the English ministers, so ready to plunge into war, were quite ignorant of its exercises; because the English people are warlike without being military, and, under the pretence of maintaining liberty which they do not possess, oppose in peace all useful martial establishments. In the beginning of each war England had to seek in blood for the knowledge necessary to insure success."

Equally has this always been the attitude of the American people towards every attempt made in peace to prepare for war. Besides this national distrust, prejudices had to be overcome which have existed both in the navy and the merchant marine. Our naval officers have never made any determined effort to create a reserve, either because they have not fully grasped the correlation and interdependence of the navy and the merchant marine, or because they have doubted the wisdom of spending upon an outside issue appropriations which, given to the navy, would produce a more immediate and tangible result. But from both points of view they are wrong:

> For a navy unsupported by a merchant marine is a hot-house plant which may produce great results for a while, but cannot endure the strain of a long protracted campaign.

From the merchant marine the *personnel* of the navy in war must come, and it is a fallacy to believe that by a small addition to our ordinary naval resources we would be able to cope with the navies of other maritime powers, or that in a long war an efficient and numerous reserve is not of greater importance than a few more seamen permanently maintained in the navy during peace.

To the merchants and ship-owners the question is one of vital importance. The earliest and most disastrous consequence of war will fall upon the shipping interest. Under any system of defence the necessities of the navy must withdraw seamen from the merchant service and raise the rate of wages. If, then, by timely precautions during peace, we can diminish the probability that war can occur at all; if we are ready upon the outbreak of war to show that our homeward-bound ships are safe; if we can abolish or modify the risk that the employment of seamen would be abruptly suspended by embargo or interfered with by impressment or draft; if we can attach the sailor to his country, and prevent him from seeking employment under other flags, surely the owners of our ships and merchants will reap the greatest advantage. Abroad the importance of the subject has been fully recognized. France, under a system which has existed for over two hundred and fifty years, maintains a reserve of 172,000 men, who are between the ages of eighteen and fifty; 65,000 of these are between the ages of twenty and twenty-six, 15,000 are usually kept afloat, and 6000 more are quartered on shore. Germany has 15,000, and England nearly the same number.

Notwithstanding the decadence of our shipping interest we have a large force from which to draw. The maritime population of this country numbers over 350,000, of whom 180,000 are available for the fleet. This number of course includes all those in any way connected with sea industries, and embraces coasters, fishermen, whalers, yachtsmen, boatmen, and all workmen in ship-building yards and equipment shops and stores.

To man our ships in time of war three means are open: voluntary enlistment, draft or impressment, or employment of men enrolled in a naval reserve. It would be unreasonable to depend altogether upon the loyal and unselfish patriotism of necessitous men serving before the mast, and there is a chance that mere enthusiasm would not induce a seaman to join the navy if employment was being offered elsewhere at increasing rates of pay. Impressment under any name is unpopular. In its common form it is illegal, and the draft is ever a last resort and always a dangerous measure. Nothing, then, remains as a certainty but to turn towards the naval reserve as the best means of manning our fleet. In time of war not only would the men enrolled come forward willingly and be immediately available, but deserters would have the machinery of the law put in motion for their apprehension, and popular feeling would be as earnest in support of their arrest as it would be opposed to all attempts which enforced the arbitrary powers of draft or impressment.

No system exists abroad entirely suited to our necessities and our national instincts; but, generally speaking, that adopted in England comes nearest to what we should employ. Naturally our lake sailors, coasters, fishermen, and yachtsmen would form the main body of the reserve. These should be enrolled, divided into classes, be given each year a certain fixed sum of pay, with an increase for each day's drill, and at stated times they should be embarked for great gun practice at sea, so they might learn something of man-of-war routine and discipline. The officers could be drawn from the merchant marine, from the graduates of the school-ships, and from former officers of the regular and volunteer services who are now in civil life.

Forced Draft

The subject of forced draft is of great importance, and, as a corollary of high-speed development, is being studied with keen interest. There are wide differences of opinion not only as to the proper systems, but even as to the value of the principle. The literature as yet is rather meagre, but an excellent compilation of existing material will be found in the latest publication of the Naval Intelligence Office. The *Marine Engineer* of September, 1887, explains:

> A forced draft in the furnaces can be generated in two ways: first, by exhausting the uptakes and funnels of the products of combustion, when a greater flow of air will necessarily take place through the fire-bars; and secondly, by increasing the pressure of the air in the furnaces beyond that of the atmosphere. The steam-blast in marine boilers is well known to engineers as a means of quickly getting up the steam after its pressure has dropped; but the locomotives on our railways afford a very good illustration of how boilers may be continuously worked under forced combustion through a jet of steam exhausting the smoke-box and funnel of the products of combustion. This system of creating a draft involves a very large expenditure of steam and water, and as it is a sine qua non in these days of high pressure that only fresh water should be used in boilers, and also as only a limited supply of this element can be carried in a ship, it follows that the plan of inducing a forced draft by means of a steam jet in the funnel cannot be well adopted in marine boilers.
>
> Mr. Martin, the inventor of the well-known furnace doors, substitutes a fan in the uptake for the steam jet, and so arranges his funnel that in the event of the forced draft not being required the gases of combustion arising from natural draft will not be impeded in their exit to the atmosphere. He claims for his invention that it does away with all necessity for closing in the stoke-holds or furnaces, and that in warships funnels could be dispensed with, as the gases and smoke could be discharged anywhere from the fans. He also claims that by his plan of producing a draft the boiler-tubes become much more efficient as heating surfaces, and that the ends of the tubes in the fire-box are not so liable to be burned away, and that therefore there will be less chance of the boiler leaking round the tubes. There appears to be some grounds for these latter assumptions, for it is

a well-known fact that the tubes of locomotive boilers, which are worked, as we have seen, on the exhaust principle, do very much more work than those of marine boilers before they are ferruled or rolled. It can also be shown by a very simple experiment that when the gases are sucked or drawn through the tubes the flame extends a much greater distance along the tube than when the gases are driven through the tubes. In this latter case the flame impinges on the tube-plates before separating into tongues and entering the tubes; but when sucked through the tongues of flame commence at some little distance from the plate before penetrating the tubes, and the ends are not therefore burned as when the flame impinges directly on them. It may be urged, however, against Martin's system that owing to the greatly increased volume of the products of combustion due to their temperature, fans of from three to four times the size of those used in other systems are required; also, that the uptakes have to be made larger and heavier to take in the fans; and lastly, that the fans themselves are likely to be quickly rendered inefficient through working in a temperature of at least a thousand degrees. These objections prove so formidable that up till the present time Martin's plan of creating a forced draft has made little or no headway.

The other plan for creating an artificial draft in marine furnaces is to force air into them by means of fans. This is done either by closing in the whole of the stoke-hold and filling it with air of a pressure greater than that of the atmosphere, or by pumping the air direct into the furnace. This latter is the usual practice in the mercantile marine, where economy of fuel is sought after. Mr. Howden seeks, by first heating the air, and then forcing it by means of fans into the furnaces and ash-pits, to insure a very rapid and complete combustion of the coal. His plan has been carried out in the Atlantic liner *Ohio* quite recently, and the results as published lead one to expect that with a little more progress in the direction in which he is working our ships will be driven across the Atlantic without the expenditure of any fuel whatever. The fact of heating the air to a temperature of two hundred degrees before it enters the furnace cannot go very far in affecting either the rapidity or the completeness of the combustion of the fuel, and it certainly cannot affect the economy. Where the fire-grate area is small compared with the total heating surface, good evaporative results are likely to be obtained; and in the *Ohio* the fire-grate area was certainly smaller than is usual for the same sized boilers fitted with forced draft. The trip of the *Ohio* to America has given somewhat different results to those of the official trials, and it is a question whether any saving in weight, either in the apparatus required to produce forced draft under this system, or in the economy of fuel to be derived from it, has been obtained more than exists in the system of closed stoke-holds.

The only plan that seems to hold its own is the closed stoke-hold system, and the results that have been obtained with it in the navy are so satisfactory that Messrs. J. & G. Thomson are about to adopt it in the two large Inman liners they are now building; and also several other firms are about to introduce it in preference to all other plans for increasing the efficiency of their boilers and promoting greater economy. In the Royal Navy space and weight are of such vital importance that the boilers have to be constructed on principles the very reverse of those which exist in boilers specially designed for high evaporative work per pound of fuel; and it is not, therefore, to be wondered at that the consumption of fuel per indicated horse-power has not been reduced since the introduction of forced draft; but, on the other hand, the capabilities of the boilers have been expanded far beyond the expectations of a few years ago. In the mercantile marine there is no reason whatever why the system of closed stoke-holds for creating a forced draft should not combine economy with greater efficiency in the boilers.

These conclusions are not universally accepted, as will be seen in the following extract from the article contributed by Assistant Engineer R. S. Griffin, United States Navy, to the Naval Intelligence Office publication mentioned at the beginning of this subject:

> The forced-draft trials of the *Archer* class go far towards sustaining the objections raised by Mr. Howden against the closed stoke-hold system. The trials of the *Archer, Brisk,* and *Cossack* had to be discontinued on several occasions, owing to leakage of the boiler-tubes; and when it is remembered that these trials are for only four hours, and that no provision is made for hoisting ashes, it becomes a question of serious consideration whether the maintenance of this high power for such a short period brings with it advantages at all comparable with the

continued development of a reasonably high power with an economical expenditure of coal, such as is possible with the closed ash-pit system.

A number of steamers have been fitted with Howden's system during the past year, among others the *Celtic*, of the White Star Line, and the *Ohio*, of the International Navigation Company, One of the latest steamers fitted with this system is the *City of Venice*, whose engines were converted from compound to quadruple expansion. Her boilers were designed to develop 1800 indicated horse-power with eighty square feet of grate, but on trial she could only work off 1300 indicated horse-power, owing to some derangement of the valves. She was afterwards tried with half the grate surface in use, when it was demonstrated that there would be no difficulty in developing the power so far as the boilers were concerned. Unfortunately, no data as to weight of boilers, .space, or heating surface are obtainable.

In 1886 the *Alliance* was supplied with new boilers, fitted with a system of forced draft designed by the Bureau of Steam Engineering. It was originally the intention of the Department to put six boilers in this vessel, as in the *Enterprise* and *Nipsic*, but with the introduction of the forced-draft system, which was purely experimental, this number was reduced to four, having a total grate surface of 128 square feet. The boilers were designed to burn anthracite coal with natural draft, and were of course unsuited to the requirements of forced draft, the ratio of heating to grate surface being only 25.6 to 1, and the water surface and steam space being small. The maximum indicated horse-power developed on trial was 1022, but any attempt to run at this or at increased power for any length of time was attended with so much priming of the boilers that the trial had to be discontinued. Alterations were made in the boilers to prevent the priming, but no continuous trial was had previously to the sailing of the *Alliance*. The results obtained on a measured base were, however, sufficient to demonstrate the practicability of the system, and to show that a higher power could be maintained with the four boilers at forced draft than with the original eight boilers at natural draft.

The practical working of the system at sea presents no difficulty, as a recent run of the *Alliance* has demonstrated. On a continuous run of ten hours, using only two boilers with sixty square feet of grate (the grate surface of each boiler having been reduced to thirty), the mean indicated horse-power was 668 and the maximum 744, being respectively 11.1 and 12.4 indicated horse-power per square foot of grate. There was an entire absence of priming, and no difficulty was experienced in operating the forced-draft apparatus, the length of the trial having been determined by the arrival of the vessel in port. The coal burned was Welsh, of fair quality, the consumption being 29.9 pounds per square foot of grate.

The efficiency of the system may be judged by the results obtained from an experimental boiler at the Washington navy-yard. The boiler was of the marine locomotive type, and had a ratio of heating to grate surface of 32.73 to 1, with a water space of 245 and a steam space of 163 cubic feet. The coal burned was ordinary Cumberland Valley bituminous, and the evaporation, when burning as much as forty pounds per square foot of grate, was 6.61, while with 37.5 it was 7.24, and this with a moderate air pressure—1.5 inches in ash-pit and one inch on furnace door.

It is unfair to attempt the explanation of this system without accompanying drawings, but it may be stated that the air, drawn by fan-blowers from the heated portion of the fire-room, is forced through a passage into the ash-pit and furnace, a portion of the current being directed by an interposed plate through the holes in the furnace frame. By the agency of a double row of holes the greater portion of the air which enters the furnace passes around the frame, thence through other apertures to the space between the furnace door and lining, and finally to the furnace through the space between the lining and furnace frame. The supply of air when firing or hauling ashes is shut off by a damper.

Appendix 2

The Question of Type

The following letter appeared in the *Times* (London) of April 4, 1885:

Sir,—May I request the favour of space in the *Times* in which to comment upon the opinions recently expressed by Sir Edward Reed and other writers respecting the designs of the *Admiral* class of ships in the Royal Navy, and the four central-citadel ships which were laid down subsequently to the *Inflexible*?

Having been closely associated with Mr. Barnaby in the designing of all these ships, with the exception of the *Ajax* and *Agamemnon*, I can speak with full knowledge of both the history and intentions of the designs.

Moreover, my share of the responsibility for the professional work involved in those designs remains, although my official connection with the constructive department of the Admiralty was severed years ago. It need hardly be added that the remarks which follow simply embody my own opinions, and that I write neither as an apologist for Mr. Barnaby nor as a champion of the ship-building policy of the Admiralty.

The sweeping condemnation which has been pronounced against the most recent English battleships is based upon the consideration of one feature only in their fighting efficiency, *viz*., the extent of the armour protection of their sides in the region of the water-line. There has been no discussion in the letters to which I have referred of the comparative speeds, armaments, or other qualities of the French and English ships. But the fact that the French ships are armour-belted from end to end, while the English ships have no vertical armour on considerable portions of the length at the region of the water-line, is considered by Sir Edward Reed so serious a matter that he says, 'The French armoured ships must in all reason be expected to dispose of these English ships in a very few minutes by simply destroying their unarmoured parts.'

From this opinion I most strongly dissent, for reasons which are stated below; and I venture to assert that if attention is directed simply to the possible effects of gunfire, while the possibly greater risks incidental to attacks with the ram and torpedo are altogether neglected, then there is ample justification for the belief that the English ships of recent design can do battle on at least equal terms with their contemporaries in the French or any other navy.

In all recent armoured ships, if the wholesale and extremely rapid destruction of the unarmoured portions of the ships which Sir Edward Reed contemplates actually took place, very considerable risks would undoubtedly result; but in my judgement these risks are not sensibly affected by the different distribution of the armour on the ships of the two great navies. And, further, there is every reason for doubting whether such wholesale destruction of the unarmoured parts could be effected with the appliances which are now available, not merely in 'a few minutes,' but in a very considerable time, and under the most favourable conditions for the attack. Nor must it be forgotten that armour, even of the greatest thickness, applied to the sides or decks of ships is not impenetrable to the attack of guns already afloat, while the *mitraille*, which is driven back into a ship when armour is penetrated, is probably as destructive as any kind of projectile can be, and at-tacks directly the vital parts which the armour is intended to protect.

In support of these assertions I must ask permission to introduce certain detailed statements which appear to be absolutely necessary to a discussion of the subject, but which shall be made as brief and untechnical as possible.

It appears that the points raised by the discussion may be grouped under two heads. First, does the shortening of the belt in the English ships introduce such serious dangers if they have to do battle with the French ships? Secondly, what should be considered the principal uses of armour-plating in modern war ships? The second question may be considered to include the first; but it will be convenient to take the questions in the order in which they have been placed, as, after all, the greatest immediate interest centres in the comparison between existing ships.

At the outset it is important to remark that in the most recent designs of armoured ships for all navies, increase in speed, armament, and thickness of armour has been associated with a decrease in the area of the broadside protected by armour. Further, it has been considered important in most cases to distribute the armoured positions of the heavy guns in the ships in order to reduce the risks of complete disablement of the principal armament by one or two lucky shots which may happen when the heavy guns are concentrated in a single citadel or battery. This distribution of the heavy guns also gives greater efficiency to the auxiliary armament of lighter guns, and enables these heavy guns to be placed at a considerably greater height above water than was usual in former days, so that the chances of the guns being prevented from being fought in heavy weather are diminished, and their power as compared with the lower guns in earlier ships is increased, especially when firing with depression.

The days of the 'completely protected ironclad,' with the broadside armoured throughout the length from the upper deck down to five or six feet below the water-line, have long gone by. The 'central battery and belt' system has also been practically dropped, whether the battery contained broadside guns or formed a citadel protecting the bases of the turrets. In short, on modern battleships there now remains only a narrow belt of armour, rising from five or six feet below the load-line to two or three feet above it. This narrow strip of armour in the French ships extends from end to end, and is associated with a protective deck worked at the height of the top of the belt, and forming a strong roof to the hold spaces beneath. In the English ships of the *Admiral* class the belt of armour extends somewhat less than half the total length, protecting one hundred and forty to one hundred and fifty feet of the central portion of the ship (in which are situate the engines and boilers), and protecting also the communications from the barbette towers to the magazines. At the extremities of the belt strong armoured bulkheads are built across the ships. The protected deck is fitted at the upper edge of the belt over the central portion. Before and abaft the bulkheads, where there is no side armour, the protection consists of a strong steel deck, situated from four to five feet below water, and extending to the bow and stern respectively. Upon this underwater deck are placed coal-bunkers, chain-lockers, fresh-water tanks, store-rooms, etc., the spaces between it and the deck next above being subdivided into a large number of water-tight compartments or cells by means of longitudinal and transverse bulkheads. A water-tight top or roof to these compartments is formed by plating over the main deck-beams with thin steel at the same height above water as the top of the armour-belt. In this manner the unarmoured ends above the protective deck are not merely packed to a large extent with water-excluding substances when the vessel is fully laden, but they are minutely subdivided into separate compartments, which can only be thrown into communication with one another by means of very extensive injuries to the partitions.

In all the modern French ships, as well as in the *Admiral* class, a light steel super-structure of considerable height is built above the level of the belt-deck; the living quarters of the crew and the stations of the auxiliary armament are contained within this light erection, which also surrounds the armoured communications from the barbette towers to the magazines. In this manner a ship with a small height of armoured freeboard is converted into a high-sided ship for all purposes of ordinary navigation, sea-worthiness, and habitability; while spaces are provided in which a more or less considerable number of light guns can be fought concurrently with the heavy guns placed in the armour-protected stations. The radical difference, therefore, between the French ships and the *Admiral* class, independently of other considerations than the armour protection of the water-line, consists in the omission of the side armour at the extremities, and the use instead of the side armour of the strong under-water deck with cellular subdivision and other arrangements for adding to the protection and securing the buoyancy of the spaces at the ends, into which water may find access through the thin sides if they are shot through and seriously damaged in action. If the completely belted French ship has to fight a vessel of the *Admiral* class, the latter has obviously the greater chance of damage to the narrow strips of the sides lying above the underwater deck before and abaft the ends of the belt. If the action takes place in smooth water, when the ships are neither rolling nor pitching, but are simply in motion, the chances of hitting these narrow strips in the water-line region might not be very great; but it must be admitted that even the lightest guns would penetrate the thin sides of the English ships and admit more or less considerable quantities of water into the ends. If, on the other hand, the fight takes place in a sea-way, with the ships rolling and pitching, then the relative importance of penetration of these narrow strips of the ends of the English ships becomes much less, because the belt armour of the French ships will be brought out of water for a considerable length of the bow and the stern by a very moderate angle of pitching, or by the passage of a comparatively low wave, and because rolling motion of the ships will alternately immerse or emerge the belt armour, even at the midships part, where it has its greatest thickness. In fact, as I have more than once said publicly, it is clearly an error to limit criticism to the longitudinal extent of the belt armour in modern ships, and to exclude consideration of the vertical extent of the armour above and below the load-line. Apart from any discussion of the question from the artillerist's point of view, or any attempt to determine the probability or otherwise of the wholesale destruction of the unarmoured portions of modern battleships by shell-fire from large guns, or by the projectiles from rapid-firing and machine guns, it is perfectly obvious to anyone who will examine into the matter that the risk of damage to the light super-structures situated above the belt must be greater than the corresponding risk of damage to the narrow strips of side area exposed at the unarmoured ends of the *Admiral* class, between the level of the belt-deck and the water-line.

Sir Spencer Robinson, after his inspection of the models shown him at the Admiralty, recognizes the fact that in the French belted ships (of which the *Amiral Duperré* is an example), if the light sides above the belt-deck are destroyed or very seriously riddled in action, the ship would be capsized in a very moderate sea-way. He further emphasizes file statement that the ships of the *Admiral* class in the English navy, if similarly treated, would also capsize under the same conditions, and he appears to be surprised at the admission having been made. The fact is that there has never been any assertion that the *Admiral* class would be safe against capsizing independently of assistance given to the armour-belted portions by the unarmoured structure situated above. On the contrary, from the first, in the design of these ships, it was recognized that their stability, in the sense of their power to resist being capsized, if inclined to even moderate angles of inclination, was not guaranteed by the armour-belts. In this respect they were in identically the same position as all other armoured ships with shallow water-line belts and isolated armoured batteries placed high above water.

What has been said respecting the Admiral class is this: If the unarmoured ends above the protective deck were completely thrown open to the sea, then the initial stability (that is to say, the stiffness of the ships or their power to resist small inclinations from the up-right) would still be guaranteed by the central armoured portions. So fully did we appreciate the fact that the life of the ship in action (as determined by her power to resist large inclinations) depends greatly upon the assistance given by the unarmoured superstructures to the armour-belted parts, that we were careful to make the structural arrangements of the superstructures above the belts such that they could bear a very considerable amount of riddling and damage from shot and shell without ceasing to contribute in the most important degree to the buoyancy and stability.

There are double cellular sides between the belt and upper decks; the main bulkheads are carried up high above water; hatches and openings are trunked up and protected by cofferdams. In short, every possible precaution is taken to subdivide into compartments, and thus limit the spaces to which water can find access when the outer sides are penetrated or shattered, as well as to facilitate the work of stopping temporarily shot-holes in the sides.

Now, without in the least intending to discredit the work of the French designers, I have to state that no corresponding or equal precautions have been taken in the portions of their ships lying above the belt-decks. And the absence of these features in the French ships is a great relative advantage to the English ships. Of course there is nothing to hinder the French from imitating our practice, but they are content to take the risks involved in a simpler construction, and in so doing they show their practical disbelief in the doctrine of armour-protected stability. I am aware that some eminent authorities do not concur with this view, and maintain that stability and buoyancy should be guaranteed by armour. To this point I will revert hereafter, but for the present I am content to say that, as between the French ships and the *Admiral* class, the most serious risks of damage by gunfire in action are of the same kind, and, practically, are not affected by the shortening of the armour-belts in the English ships.

Next I would refer to the differences which are undoubtedly involved in shortening the belts of the English ships. In the first place, by dispensing with the side armour towards the extremities a very considerable saving is effected in the weight and the cost of the armour fitted to the ships. Mr. Barnaby has recently given an illustration of this, where a ship, in other respects unchanged, has to be increased from 10,000 to 11,000 tons in displacement in order to carry the shallow armour-belt to the ends. In the *Collingwood* herself quite as large a proportionate increase of size would be involved in having a thick armour-belt from stem to stern. This saving in weight and cost of armour might, of course, be purchased too dearly, if dispensing with the armour involved possibly fatal risks to the ship. However the result may be attained, there is universal agreement that a ship-of-war should have her buoyancy, stability, and trim guaranteed as far as possible against the effects of damage in action. Now, in the *Admiral* class this matter was very carefully investigated before the design was approved. In order to prevent derangement of the trim of the vessels by penetration of the light sides above the protective deck at one end, arrangements were made in the design by means of which water can be introduced into the spaces occupied by coal-bunkers, etc., before the ships go into action.

The extent to which water may be introduced is a matter over which the captain would necessarily have control. But even if the whole of the unoccupied spaces were filled with water, the increase in draught would not exceed fourteen to eighteen inches, and the loss in speed would not exceed half a knot. If only the coal-bunkers were flooded as a preliminary to action, the chance of any serious disturbance of trim, and consequent loss of manoeuvring power or speed by damage to the light sides above the protective deck and near the water, would be very small, and the 'sinkage' of the vessel would be decreased considerably. But taking the extreme case, with the ends completely filled and a sinkage of fourteen to eighteen inches, a ship of the *Admiral* class would go into action with practically her full speed available, and with her ends so protected by under-water deck and the water admitted above that deck that damage to the thin sides by shot or shell penetrating at or near the water-line would not produce changes of trim or alterations of draught to any greater extent than would be produced if the armour-belt had been carried to the stem and stern. Nor would the admission of water into the ends render the vessel unstable.

It has been urged that the sinkage due to filling the tank ends with water is a disadvantage, because it brings the upper edge of the belt armour in the *Admiral* class about fourteen to eighteen inches nearer the water than the upper edge of the belts of the French ships. If the greatest danger of the ships was to be measured by the smallness of their 'reserve' of 'armour-protected buoyancy' (that is to say, by the buoyancy of the part of the ship lying above her fighting water-line and below the belt-deck), then the *Admiral* class would not compare favourably with the fully belted French ships. But I have already explained that this is not the true measure of the greatest danger arising from the effects of gunfire, and that it would be a, mistake to assume that in either the French or the English ships the armour-belted portions of the vessels guarantee their safety when damaged in action.

As between the *Admiral* class and the central-citadel ships of the *Inflexible* type there is a difference in this respect which has been much commented upon. When the ends of the citadel ships are filled with water, the armoured wall of the citadel still remains several feet above water; whereas, in the *Admiral* class, the top of the belt under similar conditions is very near the water-level. All that need be said on this point is that, notwithstanding the greater height of the armoured wall above water, the citadel ships have practically no greater guarantee of safety against capsizing by means of armour-protected stability than the *Admiral* class. In both classes the armoured portions require the assistance of the unarmoured to secure such a range and amount of stability as shall effectually guarantee their security when damaged in action. And, as has been stated above, this condition is true of all armour-clads with narrow armour-belts.

One other objection to the shortened belts yet remains to be considered.

It is urged that when the thin ends are broken through

or damaged by shot or shell, jagged or protruding holes will be formed in the plating near the water-line, and then if the ships are driven at speed, the water will flow into the holes in large quantities, and produce serious changes of trim and loss of speed. In support of this contention, reference is made to the published reports of experiments made with the *Inflexible*'s model about eight years ago. It is impossible to discuss the matter fully, and I must therefore content myself with a statement of my opinion, formed after a careful personal observation of these model experiments. First, it cannot be shown from the experiments that the presence of a shallow belt of armour reaching two to three feet above the still-water line would make any sensible difference in the dangers arising from the circumstances described. Holes in the thin sides above this belt would admit water in large quantities on the belt-deck when the vessel was under way, and if it could flow along that deck changes of trim, and other disagreeable consequences would result. Secondly, it is certain that the numerous bulkheads and partitions, coffer-dams, etc., built above the belt-deck level in the *Admiral* class for the very purpose of limiting the flow of entering water would greatly decrease any tendency to check the speed or change the trim. Whether the belt be short or long, it is evident that gaping holes low down in the light sides will make it prudent for a captain to slow down somewhat if he wishes to keep the water out as much as possible. But between such prudence and the danger of disaster there is a wide gulf.

Summing up the foregoing statements, I desire to record my opinion, based upon complete personal knowledge of every detail in the calculations and designs for the *Admiral* class, that the disposition of the belt-armour (In association with the protective decks and cellular sides, water-tight subdivision, etc., existing in the unarmoured portions of the vessels situated above the protective decks) is such that the buoyancy, stability, trim, speed, and manoeuvring capabilities are well guaranteed against extensive damage from shot and shell fire in action. And, further, that in these particulars the *Admiral* class are capable of meeting, at least on equal terms, their contemporary ships in the French navy.

I must add that I am not here instituting any comparison between the 'fighting efficiencies' of the ships of the two fleets;

nor have I space in this letter to do so. Opinions have differed, and will probably always differ, as to the relative importance of the different qualities which go to make up fighting efficiency. There is no simple formula admitting of general application which enables the comparative fighting values of warships to be appraised. As the conditions of naval warfare change and war material is developed, so the balance of qualities in ship designing has to be readjusted, and estimates of the fighting powers of existing ships have to be revised. And, further, different designers, working simultaneously, distribute the displacement, which is their sum total of capital to work upon, according to their own judgements of what is wisest and best for the particular conditions which the ships built from those designs have to fulfil. The designer who has the larger displacement to work upon has the better opportunity of producing a more powerful ship; but it by no means follows that he will secure so good a combination of qualities as a rival obtains on a smaller displacement. And hence I cannot but dissent from the doctrine that displacement tonnage is to be accepted as a fair measure of relative fighting efficiency, or that recent English ships are necessarily unable to fight recent French ships because they are of smaller displacement.

In the preceding remarks I have been careful to confine myself chiefly to the naval architect's side of the subject, as it would clearly be out of place for me to say much respecting the artillerist's side. But, having had the great advantage of knowing the views of some of the most experienced gun makers and gunnery officers, and having studied carefully what has been written on the subject, I would venture to say a few words.

"First, there seems, as was previously remarked, every reason for doubting, in the actual conditions of naval gunnery, whether it would be possible, not merely in a few minutes, but in a considerable time, to produce the wholesale destruction of the unarmoured parts of modern warships which has been assumed in the condemnation of the *Admiral* class. If the *Collingwood*, or one of her successors, were simply treated as a moving target in a sea-way for the *Amiral Duperré* or one of her consorts, this would be a most improbable result. But, remembering that the *Collingwood* would herself be delivering heavy blows in return for those received, the chances of her disablement would necessarily be decreased. Secondly, it does not seem at all evident that the introduction of rapid-fire guns has such an important influence on the question of shortened belts as some writers have supposed. So far as machine guns are concerned, I well remember at the board meeting which decided to approve the building of the *Collingwood* the possible effects of machine-gun fire were discussed at some length, both in reference to the adoption of the barbette system and to the system of hull protection. The rapid-firing gun which has since been introduced is now a formidable weapon; but it may be questioned whether its effects upon the unarmoured portions of modern warships would be so serious as those resulting from the shell-fire of heavier guns, and therefore it cannot with certainty be concluded that it would be advantageous to make arrangements for keeping out the projectiles from the rapid-firing guns now in use at the ends of the *Admiral* class. More especially is this true when it is considered that already rapid-fire guns of much larger calibre and greater power than the 6-pounder and 9-pounder are being made. To these guns three inches of steel would be practically no better defence than the existing thin sides, and the real defence lies in the strong protective deck. Shell-fire from heavier guns will probably be found the best form of attack against the unarmoured or lightly armoured portions of battleships, especially now that the use of steel shells with thin walls and large bursting charges is being so rapidly developed.

I would again say that on this side of the subject I do not profess to speak with authority, and it is undoubted that great differences of opinion prevail; but it must not be forgotten that the Board of Admiralty, by its recent decision announced in the House of Commons, has reaffirmed the opinion that from the artillerist's point of view the existing disposition of the armour in the *Admiral* class is satisfactory. This has been done after the attention of the Board and the public has been most strongly directed to the supposed dangers incidental to the rapid destruction of the light superstructures lying above the underwater decks of the *Admiral* class. It would be folly to suppose that in such a matter any merely personal considerations would prevent the Board from authorizing a change which was proved to be necessary or advantageous. With respect to the possibility of making experiments which should determine the points at issue, I would only say that considerable difficulties must necessarily arise in endeavouring to represent the conditions of an actual fight; but in view of the diametrically opposite views which have been expressed as to the effect of gun-fire upon cellular structures, it would certainly be advantageous if some scheme of the kind could be arranged.

There still remains to be considered the question of the uses of armour in future warships. This letter has already extended to too great a length to permit of any attempt at a full discussion. It will be admitted by all who are interested in the questions of naval design that an inquiry into the matter is urgently needed, even if it leads only to a temporary solution of the problem, in view of the present means of offence and defence.

Armour, by which term I understand not merely vertical armour, but oblique or horizontal armour, is regarded in different ways by different authorities. For example, I understand Sir Edward Reed to maintain that side-armour should be fitted in the form of a water-line belt, extending over a very considerable portion of the length, and that such armour, in association with a strong protective deck, and armoured erections for gun-stations, etc., should secure the buoyancy, trim, and stability of the vessel. At the other extreme we have the view expressed in the design of the grand Italian vessels of the *Italia* class. In them the hull-armour is only used for the purpose of assisting the cellular hull subdivisions in protecting buoyancy, stability, and trim, taking the form of a thick protective deck, which is wholly under water, and above which comes a minutely subdivided region, which Signor Brin and his colleagues consider sufficient defence against gun-fire.

In these Italian vessels the only thick armour is used to protect the gun-stations, the pilot-tower, and the communications from those important parts to the magazines and spaces below the protective deck. The strong deck, besides forming a base of the cellular subdivision, is of course a defence to the vital parts of the ship lying below it.

Between these two types of ships come the *Admiral* class of the English navy and the belted vessels of the French navy, whose resemblances and differences have been described above.

In addition, there are not a few authorities who maintain that the development of the swift torpedo-cruiser, or the swift protected cruiser, makes the continued use of armour at least questionable, seeing that to attempt to protect ships by thick armour either on decks or sides, and to secure high speeds and heavy armaments, involves the construction of large and expensive vessels, which are necessarily exposed to enormous risks in action from forms of under-water attack, against which their armour is no defence. In view of such differences of opinion, and of the heated controversies which have arisen therefrom, the time seems certainly to have arrived when some competent body should be assembled by the Admiralty for the purpose of considering the designs of our warships, and enabling our constructors to proceed with greater assurance than they can at present. Questions affecting the efficiency of the Royal Navy clearly ought not to be decided except in the most calm and dispassionate manner. The work done by the Committee on Designs for Ships of War ago was valuable, and has had important results. What is now wanted, I venture to think, is a still wider inquiry into the condition of the navy, and one of the branches of that inquiry which will require the most careful treatment is embraced in the question, 'What are the uses of armour in modern warships?'

My own opinion, reached after very careful study of the subject, is that very serious limitations have to be accepted in the disposition and general efficiency of the armaments, if the principle of protecting the stability at considerable angles of inclination by means of thick armour is accepted, the size and cost of the ships being kept within reasonable limits. There is no difficulty, of course, apart from considerations of size and cost, in fulfilling the condition of armour-protected stability; but it may be doubted whether the results could prove satisfactory, especially when the risks from underwater attacks, as well as from gunfire, are borne in mind, and the fact is recognized that even the thickest armour carried or contemplated is not proof against existing guns. No vessel can fight without running risks. It is by no means certain, however, that the greater risks to be faced are those arising from damage to the sides in the region of the water-line and consequent loss of stability. So far as I have been able to judge, it appears possible to produce a better fighting-machine for a given cost by abandoning the idea of protecting stability, buoyancy, and trim entirely by thick armour, and by the acceptance of the principle that unarmoured but specially constructed superstructures shall be trusted as contributories to the flotation and stability. Thick vertical side-armour, even over a portion of the length, appears to be by no means a necessary condition to an effective guarantee of the life and manageability of a ship when damaged in action; and it seems extremely probable that in future the great distinction between battleships and protected ships will not be found in the nature of their hull protection in the region of the water-line, but in the use of thick armour over the stations of the heavy guns in battleships.

The decisions as to future designs of our battleships is a momentous one. It can only be reached by the consideration of the relative advantages and disadvantages of alter-native proposals. It cannot be dissociated from considerations of cost for a single ship.

On all grounds, therefore, it is to be hoped that a full and impartial inquiry will be authorized without delay; for it may be assumed that, however opinions differ, there is the common desire to secure for the British navy the best types of ships and a sufficient number to insure our maritime supremacy.

I am, sir, your obedient servant,
W. H. White
Elswick Works
March 26th

The following reply by Sir Edward Reed appeared in the *Times* of April 8, 1885, the omitted portions being personal allusions which have very little bearing upon the discussion, and which are of no interest to a professional reader outside of England:

> It is not Mr. White's fault but his misfortune that he is compelled to admit the perfect correctness of the main charge which I have brought against these six ships, *viz.*, that they have been so constructed, and have been so stripped of armour protection, that their armour, even when intact and untouched, is wholly insufficient to prevent them from cap-sizing in battle. Mr. White expends a good deal of labour in attempting to show that their unarmoured parts would have a better chance of keeping the ships upright and afloat than I credit them with, which is a secondary, although an important, question; but he frankly admits that these six ships of the *Admiral* type are, and are admitted to be, so built that their 'stability in the sense of the power to resist being capsized if inclined to even moderate angles of inclination is not guaranteed by their armour-belts.'

*****

I have no doubt it would suit the purposes of all those who are or who have been responsible for those ships if I were to allow myself to be drawn, in connection with this question, away from the essential points just adverted to into a controversy upon the efforts made by the Admiralty to give to these ships, which have been denied a reasonable amount of armour protection, such relief from the grave dangers thus incurred as thin sheet compartments, coffer-dams, coals, patent fuel, stores, etc., can afford. (Cork is what was at first relied upon in this connection, but we hear no more of it now.) But I do not intend to be drawn aside from my demand for properly armoured ships of the first class by any references to these devices, and for a very simple reason, *viz.*, all such devices, whether their value be great or small, are in no sense special to armoured ships; on the contrary they are common to all ships, and are more especially applied to ships which are unable to carry armour. The application of these devices to ships stripped of armour does not make them armoured ships, any more than it makes a simple cruiser or other ordinary unarmoured vessel an armoured ship; and what I desire, and what I confidently rely upon the country demanding before long, is the construction of a few line-of-battle ships made reasonably safe by armour, in lieu of the present ships, which, while called armoured ships, in reality depend upon their thin unarmoured parts for their ability to keep upright and afloat. Besides, I do not believe in these devices for ships intended for close fighting. I even believe them likely, in not a few cases, to add to their danger rather than to their safety. If, for example, a raking shot or shell should let the sea into the compartments on one side of the ship, while those on the other side remain intact and buoyant, this very buoyancy upon the uninjured side of the ship would help to capsize her.

Mr. White says that no vessel can fight without running risks, and thinks that thick, vertical side-armour, even over a portion of the ship's length, is not a necessary guarantee of the life of a ship. Well, sir, we are all at liberty to think, or not think, what we please, so far as our sense and judgement will allow us; but Mr. White, like all other depredators of side-armour, fails utterly to show us what else there is which can be relied upon to keep shell out of a ship, or what can be done to prevent shell that burst inside a ship from spreading destruction all around. He refers us to no experiments to show that the thin plate divisions and coffer-dams, and like devices, will prove of any avail for the purpose proposed. In the absence of any such experiments, he tells us, as others have told us, that Signor Brin and colleagues in the Italian Admiralty consider 'a minutely subdivided region' at and below the water-line 'sufficient defence against gunfire.' But I do not think Signor Brin believes anything of the kind; what he believes is that the Italian government cannot afford to build a fleet of properly armoured line-of-battle ships for hard and close fighting, and that, looking at their limited resources, a few excessively fast ships, with armour here and there to protect particular parts, and with ample capabilities of retreat to a safe distance, will best serve their purpose. I do not say that he is wrong, and I certainly admire the skill which he has displayed in carrying out his well-defined object. But that object is totally different from ours, and our naval habits, our traditions, our national spirit, the very blood that flows in our veins, prevent such an object from ever becoming ours.

Mr. White all through his letter, in common with some of his late colleagues at the Admiralty, thinks and speaks as if naval warfare were henceforth to be chiefly a matter of dodging, getting chance shots, and keeping out of the enemy's way; and this may be more or less true of contests between unarmoured vessels. But why is not the line-of-battle ship *Collingwood* to be supposed to steam straight up to the enemy, I should like to know? and if she does, what is to prevent the enemy from pouring a raking fire through her bow, and ripping up at once, even with a single shell, every compartment between the stem and the transverse armoured bulkhead?

It distresses me beyond measure to see our ships constructed so as to impose upon them the most terrible penalties whenever their commanders dare, as dare they ever have, and dare they ever will, to close with their foe and try conclusions with him. Why, sir, it has been my painful duty over and over again to hear foreign officers entreat me to use all my influence against the adoption in their navy of ships with so little armoured surface as ours. On one occasion the *Collingwood* herself was imposed upon them as a model to be imitated, and I was besought to give them a safer and better ship. 'How could I ever steam up to my enemy with any confidence,' said one of the officers concerned, 'with such a ship as that under my feet?'

*****

Mr. White coolly tells us that the *Collingwood*, with five hundred tons of water logging her ends to a depth of seven or eight feet, will not be much worse off than a ship whose armoured deck stands two and a half or three feel above the water's surface, and his reason is that even above this latter deck the water would flow in when the ship was driving ahead with an injured bow. Well, sir, I will only say that sailors of experience see a very great difference between the two cases, and I can but regard such theorizings as very unfortunate basis for the designs of her majesty's ships.

I have said that Mr. White's assumptions as to the immunity of the above-water compartments and coffer-dams from widespread injury by shell-fire rest upon no experimental data; I go on to say that such data as we have to my mind point very much the other way. The *Huascar* was not an unarmoured vessel, and such shell as penetrated her had first to pass through some thin armour and wood backing; yet after the *Cochrane* and *Blanco Encaloda* had defeated her she presented internally abundant evidence of the general destruction which shell-fire produces. An officer of the *Cochrane*, who was the first person sent on board by the captors, in a letter to me written soon afterwards, said:

> It requires seeing to believe the destruction done.... We had to climb over heaps, table-high, of *débris* and dead and wounded.... We fired forty-five Palliser shell, and the engineers who were on board say that every shell, or nearly so, must have struck, and that every one that struck burst on board, doing awful destruction.

Speaking of the injury which the *Cochrane* received from a single shell of the *Huascar*, he said:

> It passed through the upper works at commander's cabin, breaking fore and aft bulkhead of cabins, breaking skylight above ward-room, athwartships bulkhead of wood, passed on, cut in two a 5-inch iron pillar, through a store-room, struck armour-plate, glanced off, passing through plating of embrasure closet at corner, finishing at after gun-port, and went overboard. This shell passed in at starboard part of stern and terminated at after battery port on port side, which is finished with the wide angle-iron, carrying out a part of the angle-iron in its flight.

This was a shell of moderate size, from a moderate gun, but it is obvious that it would have made short work of penetrating those very thin sheets of steel which constitute the compartments, coffer-dams, etc., upon the resistance of which, to my extreme surprise, those responsible for the power and safety of our fleets seem so ready to place their main dependence.

*****

For resistance to rams and torpedoes, and for the limitation of the injuries to be effected by them, as much cellular subdivision as possible should be supplied; but, as against shot and shell, subdivision by their sheet-steel is no guarantee whatever of safety in any ship, least of all in line-of-battle ships, which must be prepared for fighting at close quarters.

I must now ask for space to remark upon a few minor points in Mr. White's letter. He seems to consider that the scant armour of the *Admiral* class is somehow associated with the placing of the large, partly protected guns of these ships in separate positions, 'in order to reduce the risks of complete disablement of the principal armament by one or two lucky shots, which may happen when the heavy guns are concentrated on a single citadel or battery.' Suffice it to reply that in the proposed new designs of the Admiralty ships now before parliament, which have almost equally scant partial belts of armour, the guns are nevertheless concentrated in a single battery.

Again, Mr. White says the Admiralty have declined to adopt my advice to protect the *Admiral* class in certain unarmoured parts with 3-inch plating, and declares that such plating would practically be no better defence against rapid-fire guns than existing thin sides; but has he forgotten the fact that my suggestion has been adopted in the new designs for the protection of the battery of 6-inch guns, although it is perversely withheld from those parts of the ship in which it might assist in some degree in prolonging the ship's ability to float and to resist capsizing forces?

Mr. White makes one very singular statement. He takes exception to my claiming for the Inflexible type of ship, on account of their armoured citadel, a much better chance of retaining stability in battle than the Admiral type possesses, because, he says,' in both classes the armoured portions require the assistance of the unarmoured to secure such a range and amount of stability as shall effectually guarantee their security when damaged in action.' The fair inference to be drawn from this would be that where the principle long ago laid down by me, and supported by Mr. Barnaby in the words previously quoted, is once departed from, the danger must in all cases be so great as to exclude all distinctions of more or less risk. Mr. White can hardly mean this; but if he does not, then on what grounds are we told that a ship which has no armour at all left above water at an inclination say of six or eight degrees is no worse off than a ship which at those angles and at still greater ones has a water-light citadel over one hundred feet long to help hold her up?

* * * * *

I am not at all disposed to enter into a discussion as to the relative stabilities of the English and French ships under various conditions. The French ships have armoured belts two and a half to three feet above water from end to end. That fact, other things being presumed equal, gives them an immense advantage over our ships, which in battle trim have belts scarcely more than a foot wide above water, and for less than half their length. It is quite possible that the French constructors may have given their ships less initial stability than ours; from such information as I possess I believe they have; but in so far as the ship below the armour-deck, and the action of shot and shell upon that part of her, are concerned, whatever stability they start with in battle they will retain until their armour is pierced; whereas our ships may have a large proportion of theirs taken from them with-out their armour being pierced, and their armoured decks are then less than half the height of those of the French ships above water.

* * * * *

I will add that I doubt if the French ships are dealt fairly by at Whitehall. I lately heard a good deal of the extreme taper of their armour-belts at the bow, and the *Amiral Duperré* was always quoted in instance of this. It is true that this ship's armour does taper from fifty-five centimetres amidships to twenty-five centimetres at the stem, but she stands almost alone among recent important ships in this respect, as the following figures will show:

Name of Ship.	Thickness of Armor Amidship.	Thickness of Armor at Bows.
	Centimetres.	Centimetres.
Amiral Baudin	55	40
Formidable	55	40
Hoche	45	40
Magenta	45	40
Marceau	45	40
Caiman	50	35
Fulminant	33	25
Furieuse	50	32
Indomptable	50	37
Requin	50	40
Terrible	50	37

A friend writes me:

Comparing the *Amiral Duperré* with the *Amiral Baudin, Devastation, Formidable,* and *Foudroyant,* which are ships of about her size, the following peculiarities are observable: The *Duperré* is about three feet narrower than the other ships mentioned, and has fully fifteen inches less metacentric height. She is also slightly deeper in proportion to her breadth than the other ships.

As narrowness, small metacentric height, and excessive depth all tend to reduce stability, it would appear that the Admiralty office has, as I supposed, been careful to select a vessel not unfavourable to their purpose. But however this may be, it is no business of mine to defend the French ships in the details of their stability, nor even to defend them at all; and, as a matter of fact, the French Admiralty, although stopping far short of ours, has in my opinion gone much too far in the direction of reducing the armoured stability at considerable angles of inclination. But their falling into one error is no justification for our falling into a much greater one, and deliberately repeating it in every ship we lay down. In this connection I will only add that the experiments performed at our Admiralty on models must be viewed with great distrust for a reason not yet named. They deal only, so far as I am acquainted with them, with models set oscillating or rolling by waves or otherwise. But the danger thus dealt with is a secondary one; the primary one is that due to 'list' or prolonged inclination to one side. What sort of protection against the danger of capsizing from this cause can be possessed by a ship the entire armour on each side of which becomes immersed even in smooth water when the ship is inclined a couple of degrees only, and which then has no side left to immerse, save such as single shells can blow into holes ten by four feet?

It is to be observed that although Mr. White does not venture to join the only other apologist for these deficiently armoured ships in stating that India-rubber umbrella shot-stoppers are to be employed for their preservation in battle, he does go so far as to tell us that the spaces into which water would enter when the unarmoured parts have been penetrated have been subdivided 'to facilitate the work of stopping temporarily shot-holes in the sides,' and I know independently that a good deal of reliance is placed at the Admiralty upon the presumed ability to stop such holes as they are made. But the whole thing is a delusion. The officer of the *Cochrane*, before quoted, said:

> I wish to state that shot-plugs are out of the question after or at such a fight. They are entirely useless. Not a hole was either round, square, or oval, but different shapes—ragged, jagged, and torn, the inside parts and half-inch plating being torn in ribbons; some of the holes inside are as large as four by three feet, and of all shapes. There are many shot-plugs on board here, all sizes, conical shapes and long, but they are of no use whatever.

Mr. White's letter invites many other comments, but I have said enough to show that it in no way changes my view of the question of armour-plated line-of-battle ships. In so far as it advocates a further abandonment of armour and a further resort to doubtful devices in lieu thereof, it is already answered by anticipation by the Admiralty itself. Until I wrote my recent letters to you, our Admiralty thought as Mr. White still thinks, and tended as he still tends. In the case of all our recent cruisers but two they had abolished side-armour altogether. To my public appeal for armour-belted cruisers they have, however, responded, and are about to order six of such ships. So far, so good. We ought to be grateful for this concession to a most reasonable demand. I wish these cruisers were to be faster, much faster, but in Admiralty matters the country must be thankful for small mercies.

It only remains for me to note with satisfaction one or two of the points upon which Mr. White is in agreement with myself. He admits that it 'would certainly be advantageous' to carry out those experiments which I regard the Admiralty as afraid to make, *viz.*, experiments to test the effect of gun-fire upon the subdivided but unarmoured parts of ships.

*****

It may be taken for what it is worth, but I declare that the abandonment of armour has not at all been forced upon us by unavoidable circumstances, nor is it from any intrinsic necessity that we go on refusing to provide our ships with torpedo defence. On not immoderate dimensions, at not immoderate cost, ships might be built, still practically invulnerable to gun, ram, and torpedo alike, ships which could dispose of the *Admiral* class of ships more quickly and certainly than she could dispose of the feeblest antagonist that she is likely to encounter. But in order to produce such ships we must revive the now abandoned principle that armour, and armour alone can save from destruction those ships whose business it is to drive our future enemies from the European seas and lock them up in their own ports.

The Committee on Designs of 1872, previously alluded, to, contained sixteen members, of whom six were naval officers. Two of those members, Admiral George Elliot, R.N., and Rear-admiral A. P. Ryder, R.N., dissented so far from their colleagues that they could not sign the report, and accordingly they submitted a very able minority report embodying their views. The first of the "general principles" laid down in their report. is as follows:

> That it is of the last importance that the modifications in existing types of men-of-war which the committee have been invited to suggest should be calculated not merely to effectually meet the necessities of naval warfare now and in the immediate future, but in full view of the probable necessities of naval warfare in the more remote future.

It must be a source of satisfaction to these gallant officers to observe in some designs of the present day a confirmation of their forecast in many particulars.

The following extracts from a letter bearing upon the present controversy, by Admiral Elliot, appeared in the *Times* (London) of April 24, 1885, and contain the pith of his oft-quoted arguments:

> My first impression on reading these letters in the *Times* is one of disappointment that the point at issue between these two experts has not been more closely confined to the com-

parative merits of side-armour versus cellular-deck armour, but that their attention has been directed to this feature of design only as connected with a particular type of ship, namely, the *Collingwood*, which vessel is a hybrid, or cross between the two systems of protection to buoyancy, and therefore not truly representative of either. Mr. White's defence of the unarmoured ends of the *Collingwood* is so far unsatisfactory that it treats of a very imperfect development of the cellular-deck mode of protection, and therefore he is not an exponent of the real merits of this system.

* * * * *

I am quite aware that the main point at issue between these two distinguished naval architects has been more closely confined to the question of stability than to that of flotation as displayed in the design of the *Collingwood*, and in this scientific view of the case I do not feel competent to offer any opinion, except to point out that the cellular-deck principle *per se* does not involve any such danger as regards stability as is produced by the top weight of a central citadel. Mr. White acknowledges that this top weight will capsize his ship if deprived of the buoyancy afforded by the unarmoured ends, and on this danger point Sir Edward Reed fixes his sharpest weapon of attack.

* * * * *

The great issue at stake is how the weights available for the protection of buoyancy and for gun defence are to be distributed to the best advantage for defensive purposes, and in order to discuss Sir Edward Reed's opinions in a concise form I will deal with the question solely as concerning the use of side-armour of less than twelve inches, beyond which limit of thickness I will, for the sake of argument, admit its practical advantages; and looking to the demand for increased speed and coal-carrying capacity, it does not appear probable that if combined with adequate gun protection, and if of sufficient depth, an all-round belt of thicker than ten inches can be carried by any vessels of war except those of much greater displacement than the *Collingwood* class. I feel justified, however, in discussing the question on this basis, because Sir Edward Reed includes in his category of approved armoured ships our recent belted cruisers, having a narrow belt of ten inches maximum thickness, and takes credit for having induced the Admiralty to abandon their original intention of cellular-deck water-line protection in this class of warship in favour of this thin armour-belt.

The relative value of these two systems of water-line protection, namely, an all-round belt *versus* a raft body, must not only be ruled by the displacement decided upon for each class of vessel, and by the power of the gun which has to be encountered, but by such tactical expedients as can be resorted to in battle, as being those best suited to the known offensive and defensive properties of the combatants.

Looking at this disputed question entirely from the point of view of an artillerist and a practical seaman, I can perceive very great tactical advantages to be obtained by the adoption of the mode of protection proposed as a substitute for obsolete armour, and I view with much regret the one-sidedness of the conclusions arrived at by the opponents of this system, and the disparaging terms in which it is sought to turn it into ridicule, such as 'doubtful devices' and 'useless contrivances,' etc., because they indicate prejudice and a want of mature consideration of the incidents of naval battles. I cannot, also, help observing that while, on the one side, prophesying the most fatal consequences to ensue from what is called 'stripping ships of armour,' on the other side no admission is made of the disastrous results which must follow from placing reliance on such a delusive defensive agency as an armour-plate known to be penetrable by guns certain to be encountered; and in order to support this theory we are called upon to believe that gunners will be so excited in action or so unskilful that in no case will they hit the large object aimed at, namely, the water-line of an adversary passing even at close quarters on their beam, but I shall refer to this feature of assumed impunity hereafter.

Sir Edward Reed's comparative remarks on the effect of shot-holes as between the two systems of defence are of the same one-sided character, notwithstanding the evidence of the fractured condition of armour-plates subjected to experimental firing; and it is almost apparent that in decrying the one mode of protection he has lost sight of the fact that a ten-inch armour-plate is all that will stand between the life and death of a ship—that is to say between one well-directed shell and the magazines and boilers—which plate can be easily penetrated and smashed up by the guns which similar vessels will assuredly carry if so invited. Also, in referring to the baneful effects of raking fire and shell explosion inboard, the assumed inferiority is misplaced because one prominent advantage of the cellular-deck system is that by economizing weight at the water-line it enables the bow and stem to be armour-plated—a matter of the highest tactical importance as a defence against raking fire, which is unobtainable in a belted ship of the same displacement, at least without entailing a considerable reduction of the thickness of armour on the belt. This feature of end-on defence is not only an essential element of safety, but must prove most effective as enabling a combatant to close his adversary at an advantage, and enforce the bow-to-bow ram encounter, or compel him to resort to a stern fight, or otherwise to pass him at such close quarters as will insure direct hits and depressed fire at the water-line belt, and by these tactics the opportunities for riddling the raft body will be few and far between.

I may also express the opinion that for repairing damages in a raft-bodied ship at the water-line far more efficacious means can be resorted to than the ordinary shot-plugs, and that the use of cork bags for closing shot-holes in the coffer-dam sides, if they are open at the top, is far from being an unreasonable or ' stupid contrivance,' as it is called, considering that, as a general rule, the perforations through thin plating would not be ragged or extensive. Sir Edward Reed's wise suggestion to make the outer skin of the coffer-dam of two-inch steel plates would render machine-gun fire of little avail. The injurious effects of shell fire would, I reckon, be far more fatal if the projectile exploded in passing through the ten-inch belt than if it burst at some distance inboard after penetrating thin plating. I think it will be admitted without dispute that this feature of de-sign must be governed to a great extent by tactical considerations, the object sought for being to secure out of a given weight of steel the greatest amount of fighting vitality consistent with the power of manoeuvring available between skilful antagonists. This view of the case is especially applicable to single actions at sea, when a clever tactician will select his mode of fighting according to the offensive and defensive properties known to be possessed by his opponent, and in this respect an armour-plated bow and stern will afford enormous advantages, both for attack and defence, if the plating is extended as high as the upper deck.

In fleet actions the ram and torpedo will require more attention than the gun attack, and that feature of battle introduces another disputed point, namely, the limit of size of ship, but that question is outside the scope of the present discussion, and I shall conclude my arguments by a strong expression of opinion that, as gunpowder has so completely mastered the pretensions of outside armour protection, the direction in which prudence leans towards defensive properties in future designs for ships-of-war is that of deflection rather than of direct resistance, and that in this respect science has not reached its utmost limit of invention.

The prevailing disposition to regulate the power of the gun by the size of the vessel is, I consider, a great mistake, seeing that the additional weight of a powerful gun is not inadmissible, even in such vessels as our belted cruisers, and looking to the strong inducement held out by the continued use of armour-plating, even of such moderate thickness as ten inches. In the splendid steamers purchased from the mercantile marine, which are being armed with light guns only, one 25-ton gun would greatly add to their fighting power, but the cause of this omission may probably be found in the answer to the question. Where are the guns?

The following reply appeared in the *Times* (London) of May 1, 1885:

Sir,— The letter of Admiral Sir George Elliot ... deals ably and candidly with a subject of such fundamental importance to our navy that I venture to offer a few observations upon it.

"I am glad to see that the gallant admiral separates his case and the cellular or raft-deck system from any connection with the *Collingwood* or *Admiral* type of ship, but I regret that he has treated my criticisms of that kind of ship just as if I had applied them ill the abstract to the system which he advocates. This is not fair either to the gallant officer himself or to me, as will presently appear.

If Sir George Elliot will remove the cellular or raft-deck question completely away from the very unsatisfactory and unpleasant region of Admiralty practice, and let it be treated upon its merits, while I shall still have to respectfully submit to him some cautionary considerations, I shall also be prepared to make to him some very considerable concessions. One thing I should find it desirable to press upon him is the absolute necessity of giving closer attention to the provision of stability. He treats the subject mainly as a question of 'buoyancy,' and wisely so from his point of view; but 'stability' or the power of resisting capsizing, comes first, and on this he declines to offer an opinion. Again, when the gallant officer speaks of a 'raft' deck, I would point out that this may be a very different thing from a cellular-deck. The characteristic of a raft is that it is usually formed of solid buoyant materials; you may make it of cellular steel if you please, but in that case wherever injury lets in water the steel so far ceases to be a raft, which helps to float its load, and becomes a weight to help sink it. Now, cells formed of thin steel do not upon the face of the matter appear to be safe materials for a raft which is to be subject to the multitudinous fire of small guns and the explosions of shells of all sizes. It needs a very skilful artificer to build a safe floating raft of thin steel for such a purpose, especially when regard is had to the dangers of raking fire, against which bow and stern armour would not sufficiently provide.

Having expressed these cautions, I will go on to say that in my opinion the main idea of your gallant correspondent, which he has so long and so steadily developed, is nevertheless a sound one, and one which has a great future. I do not, of course, for a moment admit with him that the gun has yet mastered the armour. I believe the *Dreadnought*, though of old design, would still fight a good action against all ships now ready for sea, and have to fear only a very exceptional, and therefore either a very skilful or very fortunate, shot. The recent Admiralty ships, where they are armoured, are practically proof against almost every gun afloat. Further, I have satisfied myself that if the existing restrictions imposed upon us by the absence of floating docks adapted to receive ships of great breadth were removed (these restrictions crippling us to a most unfortunate degree), and if certain professional conventionalities as to the forms of ships were set aside, it would be perfectly practicable to build warships no larger and no more costly than the *Inflexible*, with enough side-armour more than a yard (three feet) thick to preserve their stability, and at the same time made ram-proof and torpedo-proof. Meanwhile, of all the vulnerable objects afloat, the recent guns themselves, by reason of their absurdly long and slender barrels, left fully exposed to all fire, are among the most vulnerable.

Still, the raft-deck system has a wide field before it, and I am quite prepared to admit that I believe in its practicability and in its sufficient security for certain classes of vessels if properly carried out. This it has not yet been in any single instance. Even in the case of the great Italian ships, as in our own, there are elements of weakness which would be fatal to the system in action, but which are not unavoidable. Allow me to assure Sir George Elliot that I have largely and closely studied this subject, and that my main objections to it are not objections of principle.

If the raft-deck system is to be adopted, it must in my opinion be carried out in a much fuller and more satisfactory manner than hitherto, and with the aid of arrangements which I have for a long time past seen the necessity of, and been engaged upon.

*****

To my mind the Admiralty, while protecting certain parts and contents of their largest ships from injury from shell fire, have made the fatal error of failing to protect the ship itself, which contains them all, from being too readily deprived of stability and made to capsize. The advocates of the alternative system must not repeat this error, or, if they do, they must not expect me to become their ally. On the other hand, if they will join me in despising what are merely specious elements of safety, and in demanding those which are real, if they will insist that our principal and most costly ships at least shall be so constructed as to keep afloat and upright for a reasonable length of time in battle, in spite of any form of attack, so as to give their gallant crews a fair chance of achieving their objects, they will not find me averse to any improvement whatever. When a suitable opportunity offers I shall be happy to show to Admiral Sir George Elliot that he has not been alone in seeking to develop the cellular or raft-deck system, and that it has, in fact, capabilities which possibly he himself may not yet have fully realized.

The same number of the *Times* contains a reply to Mr. Reed's letter of April 8, 1885, by Mr. White, mainly devoted to a refutation of certain charges of no interest to us, but containing the following paragraphs:

I must refer to the passage in which Sir Edward Reed quotes a description of the damage done to the *Huascar* in her action with the two Chilean ironclads.

This description seems to me one of the best possible illustrations of a remark in my previous letter, that 'the *mitraille* which is driven back into a ship when armour is penetrated is probably as destructive as any kind of projectile can be.' Had the *Huascar* not had weak armour, but light sides only, the local injuries might have been less. The other case cited of a shell which entered the unarmoured stern of the *Cochrane* shows how little damage may be done when a projectile passes through thin plating. At the bombardment of Alexandria there were many such examples on board our ships, although it must be frankly admitted that the engagement is no sufficient indication of what shell fire may do. A good deal of use has been made of the single case where a shell in bursting blew a hole ten by four feet in the thin sideplating of the *Superb*. The case was quite exceptional, whether it be compared with the other hits on the same ship or with the injuries done to the unarmoured sides of other ships. Moreover, in that case exceptional injury is traceable to special structural arrangements at the embrasure near the battery port, where the shell struck. These cases do not prove that the light unarmoured structures in the *Admiral* class are likely to be destroyed in such a rapid and wholesale manner as has been asserted. Nor, on the other hand, do they indicate conclusively what damage shelltire may do in future actions. On these points, as I have before remarked, experiment might be made with advantage. But, on the other hand, there is good evidence that armour so thin as to be readily penetrable to many guns may be a serious danger, and that armour over the vital parts of ships should be strong if it is to be a real defence.

* * * * *

In matters of ship design the constructors of the navy are only the servants of the Board, and while they must take sole responsibility for professional work, the governing features in the designs are determined by higher authorities, among whom are officers of large experience, both as seamen and gunners. And it is certainly not the practice of the constructive department to intrude themselves or their advice into matters for which neither their training nor their experience fits them to give an opinion.

*****

I make no attempt to be either a sailor or a gunner, but am content to seek information from the best authorities in both branches. As the result of this study of tactics and gunnery, I have been led to the belief that the sea-fights of the future are not likely to be settled altogether or chiefly by the effects of gunfire. This is not quite the same thing as Sir Edward Reed attributes to me when he says that 'Mr. White thinks and speaks as if naval warfare henceforth were to be merely a matter of dodging, getting chance shots, and keeping out of an enemy's way.'

Nor do I think that the designers of the Italian warships will indorse the description of their views and intentions, with which Sir Edward Reed has favoured us in his letter and elsewhere. I have the honour of knowing his excellency Signor Brin (later minister of marine) and other members of the constructive corps of the Italian navy, and from their statements, including the powerful publications of Signor Brin, *La Nostra Marina Militaire*, I have no hesitation in saying that in spending larger sums on single ships than have ever before been spent, the Italian authorities think, and are not alone in thinking, that they are producing the most powerful fighting ships afloat.

APPENDIX 3

Range of Guns

From Report of U. S. Fortification Board.

GUNS AFLOAT RANGING POSSIBLY NINE TO TEN MILES.

Nation.	Ship.	Maximum Armor.	Draught.	Guns.	Calibre.
		Inches.	Feet. In.	Number.	Inches.
England	Inflexible	24	25 4	4	16
France	Friedland	7⅞	29 4	2	10.6
"	Redoubtable	14	24 10	4	10.6
"	Duguesclin				
"	Bayard				
"	Turenne	9¾	24 10	4	9.5
"	Vauban				
"	Fulminant				
"	Tonnerre	13	21 4	2	10.6
Italy	Duilio	21.7	28	4	17
"	Dandolo	21.7	28 9	4	17
Germany	Sachsen				
"	Baiern				
"	Würtemberg	17.25	19 8	4	10.2
"	Baden				
"	Wespe				
"	Viper				
"	Biene				
"	Mücke				
"	Scorpion	8	10 2	1	12
"	Basilisk				
"	Cameleon				
"	Crocodil				
Brazil	Riachuelo	11	20	4	9

Besides a large number on unarmored vessels and on armored vessels not yet completed.

GUNS AFLOAT RANGING POSSIBLY TEN MILES OR UPWARD.

Nation.	Ship.	Maximum Armor.	Draught.	Guns.	Calibre.
		Inches.	Feet. In.	Number.	Inches.
England	Conqueror	12	24 0	2	12
"	Colossus	18	26 3	4	12
"	Edinburgh	18	26 3	4	12
France	Amiral Duperré	21.6	26 9	4	13.4
"	Dévastation and Foudroyant	15	24 11	{ 2 / 4	10.6 / 13.4

GUNS AFLOAT RANGING POSSIBLY TEN MILES OR UPWARD.

Nation.	Ship.	Maximum Armor.	Draught.		Guns.	Calibre.
"	Terrible	19	24	7	2	16.5
"	Tonnant	17¾	16	9	2	13.4
"	Vengeur	13¼	16	9	2	13.4
Italy	Italia	18.9	30	3	4	17
Germany	Salamander					
"	Natter	8	10	2	1	12
"	Hummel					
China	Ting Yuen	14	20		4	12
"	Chen Yuen					

GUNS RANGING POSSIBLY TEN MILES OR UPWARD SHORTLY TO BE AFLOAT.

Nation.	Ship.	Maximum Armor.	Draught.		Guns.	Calibre.
		Inches.	Feet.	In.	Number.	Inches.
England	Collingwood	18	26	3	4	12
"	Rodney	18	25	3	4	13.5
"	Benbow	18	27		2	17
"	Camperdown	18	27	3	4	13.5
"	Howe	18	27	3	4	13.5
"	Anson	18	27	3	4	13.5
"	Hero	12	24		2	12
"	Renown	18	27	3	2	16.25
"	Sanspareil	18	27	3	2	16.25
France	Amiral Baudin	21⅝	26		3	16.5
"	Formidable	21⅝	26		3	16.5
"	Furieux	19⅝	21	7	2	13.4
"	Indomptable					
"	Caïman	19⅝	24	7	2	16.5
"	Requin					
"	Marceau					
"	Hoche	17¾	27	3	2 / 2	13.4 / 10.6
"	Magenta					
"	Neptune	17¾	27	3	3	13.5
Italy	Lepanto	18.9	29	6	4	17
"	Ruggiero di Lauria	17.7	25	11	4	17
"	Andrea Doria	17.7	29	6	4	17
"	F. Morosini	17.7	25	11	4	17
Russia	Catherine II	24	27		4	12
"	Tchesme	24	25		4	12
"	Sinope	24	25		4	12
Denmark	Tordenskiold	8	15		1	13.8

ALSO FROM LEONAUR
AVAILABLE IN SOFTCOVER OR HARDCOVER WITH DUST JACKET

JOURNALS OF ROBERT ROGERS OF THE RANGERS *by Robert Rogers*—The exploits of Rogers & the Rangers in his own words during 1755-1761 in the French & Indian War.

GALLOPING GUNS *by James Young*—The Experiences of an Officer of the Bengal Horse Artillery During the Second Maratha War 1804-1805.

GORDON *by Demetrius Charles Boulger*—The Career of Gordon of Khartoum.

THE BATTLE OF NEW ORLEANS *by Zachary F. Smith*—The final major engagement of the War of 1812.

THE TWO WARS OF MRS DUBERLY *by Frances Isabella Duberly*—An Intrepid Victorian Lady's Experience of the Crimea and Indian Mutiny.

WITH THE GUARDS' BRIGADE DURING THE BOER WAR *by Edward P. Lowry*—On Campaign from Bloemfontein to Koomati Poort and Back.

THE REBELLIOUS DUCHESS *by Paul F. S. Dermoncourt*—The Adventures of the Duchess of Berri and Her Attempt to Overthrow French Monarchy.

MEN OF THE MUTINY *by John Tulloch Nash & Henry Metcalfe*—Two Accounts of the Great Indian Mutiny of 1857: Fighting with the Bengal Yeomanry Cavalry & Private Metcalfe at Lucknow.

CAMPAIGN IN THE CRIMEA *by George Shuldham Peard*—The Recollections of an Officer of the 20th Regiment of Foot.

WITHIN SEBASTOPOL *by K. Hodasevich*—A Narrative of the Campaign in the Crimea, and of the Events of the Siege.

WITH THE CAVALRY TO AFGHANISTAN *by William Taylor*—The Experiences of a Trooper of H. M. 4th Light Dragoons During the First Afghan War.

THE CAWNPORE MAN *by Mowbray Thompson*—A First Hand Account of the Siege and Massacre During the Indian Mutiny By One of Four Survivors.

BRIGADE COMMANDER: AFGHANISTAN *by Henry Brooke*—The Journal of the Commander of the 2nd Infantry Brigade, Kandahar Field Force During the Second Afghan War.

BANCROFT OF THE BENGAL HORSE ARTILLERY *by N. W. Bancroft*—An Account of the First Sikh War 1845-1846.

AVAILABLE ONLINE AT **www.leonaur.com**
AND FROM ALL GOOD BOOK STORES

ALSO FROM LEONAUR
AVAILABLE IN SOFTCOVER OR HARDCOVER WITH DUST JACKET

AFGHANISTAN: THE BELEAGUERED BRIGADE *by G. R. Gleig*—An Account of Sale's Brigade During the First Afghan War.

IN THE RANKS OF THE C. I. V *by Erskine Childers*—With the City Imperial Volunteer Battery (Honourable Artillery Company) in the Second Boer War.

THE BENGAL NATIVE ARMY *by F. G. Cardew*—An Invaluable Reference Resource.

THE 7TH (QUEEN'S OWN) HUSSARS: Volume 4—1688-1914 *by C. R. B. Barrett*—Uniforms, Equipment, Weapons, Traditions, the Services of Notable Officers and Men & the Appendices to All Volumes—Volume 4: 1688-1914.

THE SWORD OF THE CROWN *by Eric W. Sheppard*—A History of the British Army to 1914.

THE 7TH (QUEEN'S OWN) HUSSARS: Volume 3—1818-1914 *by C. R. B. Barrett*—On Campaign During the Canadian Rebellion, the Indian Mutiny, the Sudan, Matabeleland, Mashonaland and the Boer War Volume 3: 1818-1914.

THE KHARTOUM CAMPAIGN *by Bennet Burleigh*—A Special Correspondent's View of the Reconquest of the Sudan by British and Egyptian Forces under Kitchener—1898.

EL PUCHERO *by Richard McSherry*—The Letters of a Surgeon of Volunteers During Scott's Campaign of the American-Mexican War 1847-1848.

RIFLEMAN SAHIB *by E. Maude*—The Recollections of an Officer of the Bombay Rifles During the Southern Mahratta Campaign, Second Sikh War, Persian Campaign and Indian Mutiny.

THE KING'S HUSSAR *by Edwin Mole*—The Recollections of a 14th (King's) Hussar During the Victorian Era.

JOHN COMPANY'S CAVALRYMAN *by William Johnson*—The Experiences of a British Soldier in the Crimea, the Persian Campaign and the Indian Mutiny.

COLENSO & DURNFORD'S ZULU WAR *by Frances E. Colenso & Edward Durnford*—The first and possibly the most important history of the Zulu War.

U. S. DRAGOON *by Samuel E. Chamberlain*—Experiences in the Mexican War 1846-48 and on the South Western Frontier.

AVAILABLE ONLINE AT **www.leonaur.com**
AND FROM ALL GOOD BOOK STORES

ALSO FROM LEONAUR
AVAILABLE IN SOFTCOVER OR HARDCOVER WITH DUST JACKET

THE 2ND MAORI WAR: 1860-1861 by Robert Carey—The Second Maori War, or First Taranaki War, one more bloody instalment of the conflicts between European settlers and the indigenous Maori people.

A JOURNAL OF THE SECOND SIKH WAR by Daniel A. Sandford—The Experiences of an Ensign of the 2nd Bengal European Regiment During the Campaign in the Punjab, India, 1848-49.

THE LIGHT INFANTRY OFFICER by John H. Cooke—The Experiences of an Officer of the 43rd Light Infantry in America During the War of 1812.

BUSHVELDT CARBINEERS by George Witton—The War Against the Boers in South Africa and the 'Breaker' Morant Incident.

LAKE'S CAMPAIGNS IN INDIA by Hugh Pearse—The Second Anglo Maratha War, 1803-1807.

BRITAIN IN AFGHANISTAN 1: THE FIRST AFGHAN WAR 1839-42 by Archibald Forbes—From invasion to destruction-a British military disaster.

BRITAIN IN AFGHANISTAN 2: THE SECOND AFGHAN WAR 1878-80 by Archibald Forbes—This is the history of the Second Afghan War-another episode of British military history typified by savagery, massacre, siege and battles.

UP AMONG THE PANDIES by Vivian Dering Majendie—Experiences of a British Officer on Campaign During the Indian Mutiny, 1857-1858.

MUTINY: 1857 by James Humphries—Authentic Voices from the Indian Mutiny-First Hand Accounts of Battles, Sieges and Personal Hardships.

BLOW THE BUGLE, DRAW THE SWORD by W. H. G. Kingston—The Wars, Campaigns, Regiments and Soldiers of the British & Indian Armies During the Victorian Era, 1839-1898.

WAR BEYOND THE DRAGON PAGODA by Major J. J. Snodgrass—A Personal Narrative of the First Anglo-Burmese War 1824 - 1826.

THE HERO OF ALIWAL by James Humphries—The Campaigns of Sir Harry Smith in India, 1843-1846, During the Gwalior War & the First Sikh War.

ALL FOR A SHILLING A DAY by Donald F. Featherstone—The story of H.M. 16th, the Queen's Lancers During the first Sikh War 1845-1846.

AVAILABLE ONLINE AT **www.leonaur.com**
AND FROM ALL GOOD BOOK STORES

ALSO FROM LEONAUR
AVAILABLE IN SOFTCOVER OR HARDCOVER WITH DUST JACKET

THE FALL OF THE MOGHUL EMPIRE OF HINDUSTAN by H. G. Keene—By the beginning of the nineteenth century, as British and Indian armies under Lake and Wellesley dominated the scene, a little over half a century of conflict brought the Moghul Empire to its knees.

LADY SALE'S AFGHANISTAN by Florentia Sale—An Indomitable Victorian Lady's Account of the Retreat from Kabul During the First Afghan War.

THE CAMPAIGN OF MAGENTA AND SOLFERINO 1859 by Harold Carmichael Wylly—The Decisive Conflict for the Unification of Italy.

FRENCH'S CAVALRY CAMPAIGN by J. G. Maydon—A Special Correspondent's View of British Army Mounted Troops During the Boer War.

CAVALRY AT WATERLOO by Sir Evelyn Wood—British Mounted Troops During the Campaign of 1815.

THE SUBALTERN by George Robert Gleig—The Experiences of an Officer of the 85th Light Infantry During the Peninsular War.

NAPOLEON AT BAY, 1814 by F. Loraine Petre—The Campaigns to the Fall of the First Empire.

NAPOLEON AND THE CAMPAIGN OF 1806 by Colonel Vachée—The Napoleonic Method of Organisation and Command to the Battles of Jena & Auerstädt.

THE COMPLETE ADVENTURES IN THE CONNAUGHT RANGERS by William Grattan—The 88th Regiment during the Napoleonic Wars by a Serving Officer.

BUGLER AND OFFICER OF THE RIFLES by William Green & Harry Smith—With the 95th (Rifles) during the Peninsular & Waterloo Campaigns of the Napoleonic Wars.

NAPOLEONIC WAR STORIES by Sir Arthur Quiller-Couch—Tales of soldiers, spies, battles & sieges from the Peninsular & Waterloo campaingns.

CAPTAIN OF THE 95TH (RIFLES) by Jonathan Leach—An officer of Wellington's sharpshooters during the Peninsular, South of France and Waterloo campaigns of the Napoleonic wars.

RIFLEMAN COSTELLO by Edward Costello—The adventures of a soldier of the 95th (Rifles) in the Peninsular & Waterloo Campaigns of the Napoleonic wars.

AVAILABLE ONLINE AT www.leonaur.com
AND FROM ALL GOOD BOOK STORES

ALSO FROM LEONAUR
AVAILABLE IN SOFTCOVER OR HARDCOVER WITH DUST JACKET

AT THEM WITH THE BAYONET by *Donald F. Featherstone*—The first Anglo-Sikh War 1845-1846.

STEPHEN CRANE'S BATTLES by *Stephen Crane*—Nine Decisive Battles Recounted by the Author of 'The Red Badge of Courage'.

THE GURKHA WAR by *H. T. Prinsep*—The Anglo-Nepalese Conflict in North East India 1814-1816.

FIRE & BLOOD by *G. R. Gleig*—The burning of Washington & the battle of New Orleans, 1814, through the eyes of a young British soldier.

SOUND ADVANCE! by *Joseph Anderson*—Experiences of an officer of HM 50th regiment in Australia, Burma & the Gwalior war.

THE CAMPAIGN OF THE INDUS by *Thomas Holdsworth*—Experiences of a British Officer of the 2nd (Queen's Royal) Regiment in the Campaign to Place Shah Shuja on the Throne of Afghanistan 1838 - 1840.

WITH THE MADRAS EUROPEAN REGIMENT IN BURMA by *John Butler*—The Experiences of an Officer of the Honourable East India Company's Army During the First Anglo-Burmese War 1824 - 1826.

IN ZULULAND WITH THE BRITISH ARMY by *Charles L. Norris-Newman*—The Anglo-Zulu war of 1879 through the first-hand experiences of a special correspondent.

BESIEGED IN LUCKNOW by *Martin Richard Gubbins*—The first Anglo-Sikh War 1845-1846.

A TIGER ON HORSEBACK by *L. March Phillips*—The Experiences of a Trooper & Officer of Rimington's Guides - The Tigers - during the Anglo-Boer war 1899 - 1902.

SEPOYS, SIEGE & STORM by *Charles John Griffiths*—The Experiences of a young officer of H.M.'s 61st Regiment at Ferozepore, Delhi ridge and at the fall of Delhi during the Indian mutiny 1857.

CAMPAIGNING IN ZULULAND by *W. E. Montague*—Experiences on campaign during the Zulu war of 1879 with the 94th Regiment.

THE STORY OF THE GUIDES by *G.J. Younghusband*—The Exploits of the Soldiers of the famous Indian Army Regiment from the northwest frontier 1847 - 1900.

AVAILABLE ONLINE AT **www.leonaur.com**
AND FROM ALL GOOD BOOK STORES

ALSO FROM LEONAUR
AVAILABLE IN SOFTCOVER OR HARDCOVER WITH DUST JACKET

ZULU:1879 *by D.C.F. Moodie & the Leonaur Editors*—The Anglo-Zulu War of 1879 from contemporary sources: First Hand Accounts, Interviews, Dispatches, Official Documents & Newspaper Reports.

THE RED DRAGOON *by W.J. Adams*—With the 7th Dragoon Guards in the Cape of Good Hope against the Boers & the Kaffir tribes during the 'war of the axe' 1843-48'.

THE RECOLLECTIONS OF SKINNER OF SKINNER'S HORSE *by James Skinner*—James Skinner and his 'Yellow Boys' Irregular cavalry in the wars of India between the British, Mahratta, Rajput, Mogul, Sikh & Pindarree Forces.

A CAVALRY OFFICER DURING THE SEPOY REVOLT *by A. R. D. Mackenzie*—Experiences with the 3rd Bengal Light Cavalry, the Guides and Sikh Irregular Cavalry from the outbreak to Delhi and Lucknow.

A NORFOLK SOLDIER IN THE FIRST SIKH WAR *by J W Baldwin*—Experiences of a private of H.M. 9th Regiment of Foot in the battles for the Punjab, India 1845-6.

TOMMY ATKINS' WAR STORIES: 14 FIRST HAND ACCOUNTS—Fourteen first hand accounts from the ranks of the British Army during Queen Victoria's Empire.

THE WATERLOO LETTERS *by H. T. Siborne*—Accounts of the Battle by British Officers for its Foremost Historian.

NEY: GENERAL OF CAVALRY VOLUME 1—1769-1799 *by Antoine Bulos*—The Early Career of a Marshal of the First Empire.

NEY: MARSHAL OF FRANCE VOLUME 2—1799-1805 *by Antoine Bulos*—The Early Career of a Marshal of the First Empire.

AIDE-DE-CAMP TO NAPOLEON *by Philippe-Paul de Ségur*—For anyone interested in the Napoleonic Wars this book, written by one who was intimate with the strategies and machinations of the Emperor, will be essential reading.

TWILIGHT OF EMPIRE *by Sir Thomas Ussher & Sir George Cockburn*—Two accounts of Napoleon's Journeys in Exile to Elba and St. Helena: Narrative of Events by Sir Thomas Ussher & Napoleon's Last Voyage: Extract of a diary by Sir George Cockburn.

PRIVATE WHEELER *by William Wheeler*—The letters of a soldier of the 51st Light Infantry during the Peninsular War & at Waterloo.

AVAILABLE ONLINE AT **www.leonaur.com**
AND FROM ALL GOOD BOOK STORES

ALSO FROM LEONAUR
AVAILABLE IN SOFTCOVER OR HARDCOVER WITH DUST JACKET

OFFICERS & GENTLEMEN *by Peter Hawker & William Graham*—Two Accounts of British Officers During the Peninsula War: Officer of Light Dragoons by Peter Hawker & Campaign in Portugal and Spain by William Graham .

THE WALCHEREN EXPEDITION *by Anonymous*—The Experiences of a British Officer of the 81st Regt. During the Campaign in the Low Countries of 1809.

LADIES OF WATERLOO *by Charlotte A. Eaton, Magdalene de Lancey & Juana Smith*—The Experiences of Three Women During the Campaign of 1815: Waterloo Days by Charlotte A. Eaton, A Week at Waterloo by Magdalene de Lancey & Juana's Story by Juana Smith.

JOURNAL OF AN OFFICER IN THE KING'S GERMAN LEGION *by John Frederick Hering*—Recollections of Campaigning During the Napoleonic Wars.

JOURNAL OF AN ARMY SURGEON IN THE PENINSULAR WAR *by Charles Boutflower*—The Recollections of a British Army Medical Man on Campaign During the Napoleonic Wars.

ON CAMPAIGN WITH MOORE AND WELLINGTON *by Anthony Hamilton*—The Experiences of a Soldier of the 43rd Regiment During the Peninsular War.

THE ROAD TO AUSTERLITZ *by R. G. Burton*—Napoleon's Campaign of 1805.

SOLDIERS OF NAPOLEON *by A. J. Doisy De Villargennes & Arthur Chuquet*—The Experiences of the Men of the French First Empire: Under the Eagles by A. J. Doisy De Villargennes & Voices of 1812 by Arthur Chuquet .

INVASION OF FRANCE, 1814 *by F. W. O. Maycock*—The Final Battles of the Napoleonic First Empire.

LEIPZIG—A CONFLICT OF TITANS *by Frederic Shoberl*—A Personal Experience of the 'Battle of the Nations' During the Napoleonic Wars, October 14th-19th, 1813.

SLASHERS *by Charles Cadell*—The Campaigns of the 28th Regiment of Foot During the Napoleonic Wars by a Serving Officer.

BATTLE IMPERIAL *by Charles William Vane*—The Campaigns in Germany & France for the Defeat of Napoleon 1813-1814.

SWIFT & BOLD *by Gibbes Rigaud*—The 60th Rifles During the Peninsula War.

AVAILABLE ONLINE AT **www.leonaur.com**
AND FROM ALL GOOD BOOK STORES

ALSO FROM LEONAUR
AVAILABLE IN SOFTCOVER OR HARDCOVER WITH DUST JACKET

ADVENTURES OF A YOUNG RIFLEMAN by *Johann Christian Maempel*—The Experiences of a Saxon in the French & British Armies During the Napoleonic Wars.

THE HUSSAR by *Norbert Landsheit & G. R. Gleig*—A German Cavalryman in British Service Throughout the Napoleonic Wars.

RECOLLECTIONS OF THE PENINSULA by *Moyle Sherer*—An Officer of the 34th Regiment of Foot—'The Cumberland Gentlemen'—on Campaign Against Napoleon's French Army in Spain.

MARINE OF REVOLUTION & CONSULATE by *Moreau de Jonnès*—The Recollections of a French Soldier of the Revolutionary Wars 1791-1804.

GENTLEMEN IN RED by *John Dobbs & Robert Knowles*—Two Accounts of British Infantry Officers During the Peninsular War Recollections of an Old 52nd Man by John Dobbs An Officer of Fusiliers by Robert Knowles.

CORPORAL BROWN'S CAMPAIGNS IN THE LOW COUNTRIES by *Robert Brown*—Recollections of a Coldstream Guard in the Early Campaigns Against Revolutionary France 1793-1795.

THE 7TH (QUEENS OWN) HUSSARS: Volume 2—1793-1815 by *C. R. B. Barrett*—During the Campaigns in the Low Countries & the Peninsula and Waterloo Campaigns of the Napoleonic Wars. Volume 2: 1793-1815.

THE MARENGO CAMPAIGN 1800 by *Herbert H. Sargent*—The Victory that Completed the Austrian Defeat in Italy.

DONALDSON OF THE 94TH—SCOTS BRIGADE by *Joseph Donaldson*—The Recollections of a Soldier During the Peninsula & South of France Campaigns of the Napoleonic Wars.

A CONSCRIPT FOR EMPIRE by *Philippe as told to Johann Christian Maempel*—The Experiences of a Young German Conscript During the Napoleonic Wars.

JOURNAL OF THE CAMPAIGN OF 1815 by *Alexander Cavalié Mercer*—The Experiences of an Officer of the Royal Horse Artillery During the Waterloo Campaign.

NAPOLEON'S CAMPAIGNS IN POLAND 1806-7 by *Robert Wilson*—The campaign in Poland from the Russian side of the conflict.

AVAILABLE ONLINE AT **www.leonaur.com**
AND FROM ALL GOOD BOOK STORES

ALSO FROM LEONAUR
AVAILABLE IN SOFTCOVER OR HARDCOVER WITH DUST JACKET

OMPTEDA OF THE KING'S GERMAN LEGION by *Christian von Ompteda*—A Hanoverian Officer on Campaign Against Napoleon.

LIEUTENANT SIMMONS OF THE 95TH (RIFLES) by *George Simmons*—Recollections of the Peninsula, South of France & Waterloo Campaigns of the Napoleonic Wars.

A HORSEMAN FOR THE EMPEROR by *Jean Baptiste Gazzola*—A Cavalryman of Napoleon's Army on Campaign Throughout the Napoleonic Wars.

SERGEANT LAWRENCE by *William Lawrence*—With the 40th Regt. of Foot in South America, the Peninsular War & at Waterloo.

CAMPAIGNS WITH THE FIELD TRAIN by *Richard D. Henegan*—Experiences of a British Officer During the Peninsula and Waterloo Campaigns of the Napoleonic Wars.

CAVALRY SURGEON by *S. D. Broughton*—On Campaign Against Napoleon in the Peninsula & South of France During the Napoleonic Wars 1812-1814.

MEN OF THE RIFLES by *Thomas Knight, Henry Curling & Jonathan Leach*—The Reminiscences of Thomas Knight of the 95th (Rifles) by Thomas Knight, Henry Curling's Anecdotes by Henry Curling & The Field Services of the Rifle Brigade from its Formation to Waterloo by Jonathan Leach.

THE ULM CAMPAIGN 1805 by *F. N. Maude*—Napoleon and the Defeat of the Austrian Army During the 'War of the Third Coalition'.

SOLDIERING WITH THE 'DIVISION' by *Thomas Garrety*—The Military Experiences of an Infantryman of the 43rd Regiment During the Napoleonic Wars.

SERGEANT MORRIS OF THE 73RD FOOT by *Thomas Morris*—The Experiences of a British Infantryman During the Napoleonic Wars-Including Campaigns in Germany and at Waterloo.

A VOICE FROM WATERLOO by *Edward Cotton*—The Personal Experiences of a British Cavalryman Who Became a Battlefield Guide and Authority on the Campaign of 1815.

NAPOLEON AND HIS MARSHALS by *J. T. Headley*—The Men of the First Empire.

AVAILABLE ONLINE AT **www.leonaur.com**
AND FROM ALL GOOD BOOK STORES

ALSO FROM LEONAUR

AVAILABLE IN SOFTCOVER OR HARDCOVER WITH DUST JACKET

COLBORNE: A SINGULAR TALENT FOR WAR by *John Colborne*—The Napoleonic Wars Career of One of Wellington's Most Highly Valued Officers in Egypt, Holland, Italy, the Peninsula and at Waterloo.

NAPOLEON'S RUSSIAN CAMPAIGN by *Philippe Henri de Segur*—The Invasion, Battles and Retreat by an Aide-de-Camp on the Emperor's Staff.

WITH THE LIGHT DIVISION by *John H. Cooke*—The Experiences of an Officer of the 43rd Light Infantry in the Peninsula and South of France During the Napoleonic Wars.

WELLINGTON AND THE PYRENEES CAMPAIGN VOLUME I: FROM VITORIA TO THE BIDASSOA by *F. C. Beatson*—The final phase of the campaign in the Iberian Peninsula.

WELLINGTON AND THE INVASION OF FRANCE VOLUME II: THE BIDASSOA TO THE BATTLE OF THE NIVELLE by *F. C. Beatson*—The final phase of the campaign in the Iberian Peninsula.

WELLINGTON AND THE FALL OF FRANCE VOLUME III: THE GAVES AND THE BATTLE OF ORTHEZ by *F. C. Beatson*—The final phase of the campaign in the Iberian Peninsula.

NAPOLEON'S IMPERIAL GUARD: FROM MARENGO TO WATERLOO by *J. T. Headley*—The story of Napoleon's Imperial Guard and the men who commanded them.

BATTLES & SIEGES OF THE PENINSULAR WAR by *W. H. Fitchett*—Corunna, Busaco, Albuera, Ciudad Rodrigo, Badajos, Salamanca, San Sebastian & Others.

SERGEANT GUILLEMARD: THE MAN WHO SHOT NELSON? by *Robert Guillemard*—A Soldier of the Infantry of the French Army of Napoleon on Campaign Throughout Europe.

WITH THE GUARDS ACROSS THE PYRENEES by *Robert Batty*—The Experiences of a British Officer of Wellington's Army During the Battles for the Fall of Napoleonic France, 1813 .

A STAFF OFFICER IN THE PENINSULA by *E. W. Buckham*—An Officer of the British Staff Corps Cavalry During the Peninsula Campaign of the Napoleonic Wars.

THE LEIPZIG CAMPAIGN: 1813—NAPOLEON AND THE "BATTLE OF THE NATIONS" by *F. N. Maude*—Colonel Maude's analysis of Napoleon's campaign of 1813 around Leipzig.

AVAILABLE ONLINE AT **www.leonaur.com**
AND FROM ALL GOOD BOOK STORES

ALSO FROM LEONAUR
AVAILABLE IN SOFTCOVER OR HARDCOVER WITH DUST JACKET

BUGEAUD: A PACK WITH A BATON by *Thomas Robert Bugeaud*—The Early Campaigns of a Soldier of Napoleon's Army Who Would Become a Marshal of France.

WATERLOO RECOLLECTIONS by *Frederick Llewellyn*—Rare First Hand Accounts, Letters, Reports and Retellings from the Campaign of 1815.

SERGEANT NICOL by *Daniel Nicol*—The Experiences of a Gordon Highlander During the Napoleonic Wars in Egypt, the Peninsula and France.

THE JENA CAMPAIGN: 1806 by *F. N. Maude*—The Twin Battles of Jena & Auerstadt Between Napoleon's French and the Prussian Army.

PRIVATE O'NEIL by *Charles O'Neil*—The recollections of an Irish Rogue of H. M. 28th Regt.—The Slashers—during the Peninsula & Waterloo campaigns of the Napoleonic war.

ROYAL HIGHLANDER by *James Anton*—A soldier of H.M 42nd (Royal) Highlanders during the Peninsular, South of France & Waterloo Campaigns of the Napoleonic Wars.

CAPTAIN BLAZE by *Elzéar Blaze*—Life in Napoleons Army.

LEJEUNE VOLUME 1 by *Louis-François Lejeune*—The Napoleonic Wars through the Experiences of an Officer on Berthier's Staff.

LEJEUNE VOLUME 2 by *Louis-François Lejeune*—The Napoleonic Wars through the Experiences of an Officer on Berthier's Staff.

CAPTAIN COIGNET by *Jean-Roch Coignet*—A Soldier of Napoleon's Imperial Guard from the Italian Campaign to Russia and Waterloo.

FUSILIER COOPER by *John S. Cooper*—Experiences in the 7th (Royal) Fusiliers During the Peninsular Campaign of the Napoleonic Wars and the American Campaign to New Orleans.

FIGHTING NAPOLEON'S EMPIRE by *Joseph Anderson*—The Campaigns of a British Infantryman in Italy, Egypt, the Peninsular & the West Indies During the Napoleonic Wars.

CHASSEUR BARRES by *Jean-Baptiste Barres*—The experiences of a French Infantryman of the Imperial Guard at Austerlitz, Jena, Eylau, Friedland, in the Peninsular, Lutzen, Bautzen, Zinnwald and Hanau during the Napoleonic Wars.

AVAILABLE ONLINE AT **www.leonaur.com**
AND FROM ALL GOOD BOOK STORES

ALSO FROM LEONAUR
AVAILABLE IN SOFTCOVER OR HARDCOVER WITH DUST JACKET

CAPTAIN COIGNET *by Jean-Roch Coignet*—A Soldier of Napoleon's Imperial Guard from the Italian Campaign to Russia and Waterloo.

HUSSAR ROCCA *by Albert Jean Michel de Rocca*—A French cavalry officer's experiences of the Napoleonic Wars and his views on the Peninsular Campaigns against the Spanish, British And Guerilla Armies.

MARINES TO 95TH (RIFLES) *by Thomas Fernyhough*—The military experiences of Robert Fernyhough during the Napoleonic Wars.

LIGHT BOB *by Robert Blakeney*—The experiences of a young officer in H.M 28th & 36th regiments of the British Infantry during the Peninsular Campaign of the Napoleonic Wars 1804 - 1814.

WITH WELLINGTON'S LIGHT CAVALRY *by William Tomkinson*—The Experiences of an officer of the 16th Light Dragoons in the Peninsular and Waterloo campaigns of the Napoleonic Wars.

SERGEANT BOURGOGNE *by Adrien Bourgogne*—With Napoleon's Imperial Guard in the Russian Campaign and on the Retreat from Moscow 1812 - 13.

SURTEES OF THE 95TH (RIFLES) *by William Surtees*—A Soldier of the 95th (Rifles) in the Peninsular campaign of the Napoleonic Wars.

SWORDS OF HONOUR *by Henry Newbolt & Stanley L. Wood*—The Careers of Six Outstanding Officers from the Napoleonic Wars, the Wars for India and the American Civil War.

ENSIGN BELL IN THE PENINSULAR WAR *by George Bell*—The Experiences of a young British Soldier of the 34th Regiment 'The Cumberland Gentlemen' in the Napoleonic wars.

HUSSAR IN WINTER *by Alexander Gordon*—A British Cavalry Officer during the retreat to Corunna in the Peninsular campaign of the Napoleonic Wars.

THE COMPLEAT RIFLEMAN HARRIS *by Benjamin Harris as told to and transcribed by Captain Henry Curling, 52nd Regt. of Foot*—The adventures of a soldier of the 95th (Rifles) during the Peninsular Campaign of the Napoleonic Wars.

THE ADVENTURES OF A LIGHT DRAGOON *by George Farmer & G.R. Gleig*—A cavalryman during the Peninsular & Waterloo Campaigns, in captivity & at the siege of Bhurtpore, India.

AVAILABLE ONLINE AT **www.leonaur.com**
AND FROM ALL GOOD BOOK STORES

ALSO FROM LEONAUR
AVAILABLE IN SOFTCOVER OR HARDCOVER WITH DUST JACKET

THE LIFE OF THE REAL BRIGADIER GERARD VOLUME 1—THE YOUNG HUSSAR 1782-1807 *by Jean-Baptiste De Marbot*—A French Cavalryman Of the Napoleonic Wars at Marengo, Austerlitz, Jena, Eylau & Friedland.

THE LIFE OF THE REAL BRIGADIER GERARD VOLUME 2—IMPERIAL AIDE-DE-CAMP 1807-1811 *by Jean-Baptiste De Marbot*—A French Cavalryman of the Napoleonic Wars at Saragossa, Landshut, Eckmuhl, Ratisbon, Aspern-Essling, Wagram, Busaco & Torres Vedras.

THE LIFE OF THE REAL BRIGADIER GERARD VOLUME 3—COLONEL OF CHASSEURS 1811-1815 *by Jean-Baptiste De Marbot*—A French Cavalryman in the retreat from Moscow, Lutzen, Bautzen, Katzbach, Leipzig, Hanau & Waterloo.

THE INDIAN WAR OF 1864 *by Eugene Ware*—The Experiences of a Young Officer of the 7th Iowa Cavalry on the Western Frontier During the Civil War.

THE MARCH OF DESTINY *by Charles E. Young & V. Devinny*—Dangers of the Trail in 1865 by Charles E. Young & The Story of a Pioneer by V. Devinny, two Accounts of Early Emigrants to Colorado.

CROSSING THE PLAINS *by William Audley Maxwell*—A First Hand Narrative of the Early Pioneer Trail to California in 1857.

CHIEF OF SCOUTS *by William F. Drannan*—A Pilot to Emigrant and Government Trains, Across the Plains of the Western Frontier.

THIRTY-ONE YEARS ON THE PLAINS AND IN THE MOUNTAINS *by William F. Drannan*—William Drannan was born to be a pioneer, hunter, trapper and wagon train guide during the momentous days of the Great American West.

THE INDIAN WARS VOLUNTEER *by William Thompson*—Recollections of the Conflict Against the Snakes, Shoshone, Bannocks, Modocs and Other Native Tribes of the American North West.

THE 4TH TENNESSEE CAVALRY *by George B. Guild*—The Services of Smith's Regiment of Confederate Cavalry by One of its Officers.

COLONEL WORTHINGTON'S SHILOH *by T. Worthington*—The Tennessee Campaign, 1862, by an Officer of the Ohio Volunteers.

FOUR YEARS IN THE SADDLE *by W. L. Curry*—The History of the First Regiment Ohio Volunteer Cavalry in the American Civil War.

AVAILABLE ONLINE AT **www.leonaur.com**
AND FROM ALL GOOD BOOK STORES

ALSO FROM LEONAUR
AVAILABLE IN SOFTCOVER OR HARDCOVER WITH DUST JACKET

LIFE IN THE ARMY OF NORTHERN VIRGINIA *by Carlton McCarthy*—The Observations of a Confederate Artilleryman of Cutshaw's Battalion During the American Civil War 1861-1865.

HISTORY OF THE CAVALRY OF THE ARMY OF THE POTOMAC *by Charles D. Rhodes*—Including Pope's Army of Virginia and the Cavalry Operations in West Virginia During the American Civil War.

CAMP-FIRE AND COTTON-FIELD *by Thomas W. Knox*—A New York Herald Correspondent's View of the American Civil War.

SERGEANT STILLWELL *by Leander Stillwell*—The Experiences of a Union Army Soldier of the 61st Illinois Infantry During the American Civil War.

STONEWALL'S CANNONEER *by Edward A. Moore*—Experiences with the Rockbridge Artillery, Confederate Army of Northern Virginia, During the American Civil War.

THE SIXTH CORPS *by George Stevens*—The Army of the Potomac, Union Army, During the American Civil War.

THE RAILROAD RAIDERS *by William Pittenger*—An Ohio Volunteers Recollections of the Andrews Raid to Disrupt the Confederate Railroad in Georgia During the American Civil War.

CITIZEN SOLDIER *by John Beatty*—An Account of the American Civil War by a Union Infantry Officer of Ohio Volunteers Who Became a Brigadier General.

COX: PERSONAL RECOLLECTIONS OF THE CIVIL WAR--VOLUME 1 *by Jacob Dolson Cox*—West Virginia, Kanawha Valley, Gauley Bridge, Cotton Mountain, South Mountain, Antietam, the Morgan Raid & the East Tennessee Campaign.

COX: PERSONAL RECOLLECTIONS OF THE CIVIL WAR--VOLUME 2 *by Jacob Dolson Cox*—Siege of Knoxville, East Tennessee, Atlanta Campaign, the Nashville Campaign & the North Carolina Campaign.

KERSHAW'S BRIGADE VOLUME 1 *by D. Augustus Dickert*—Manassas, Seven Pines, Sharpsburg (Antietam), Fredricksburg, Chancellorsville, Gettysburg, Chickamauga, Chattanooga, Fort Sanders & Bean Station.

KERSHAW'S BRIGADE VOLUME 2 *by D. Augustus Dickert*—At the wilderness, Cold Harbour, Petersburg, The Shenandoah Valley and Cedar Creek..

AVAILABLE ONLINE AT **www.leonaur.com**
AND FROM ALL GOOD BOOK STORES

ALSO FROM LEONAUR
AVAILABLE IN SOFTCOVER OR HARDCOVER WITH DUST JACKET

THE RELUCTANT REBEL by *William G. Stevenson*—A young Kentuckian's experiences in the Confederate Infantry & Cavalry during the American Civil War..

BOOTS AND SADDLES by *Elizabeth B. Custer*—The experiences of General Custer's Wife on the Western Plains.

FANNIE BEERS' CIVIL WAR by *Fannie A. Beers*—A Confederate Lady's Experiences of Nursing During the Campaigns & Battles of the American Civil War.

LADY SALE'S AFGHANISTAN by *Florentia Sale*—An Indomitable Victorian Lady's Account of the Retreat from Kabul During the First Afghan War.

THE TWO WARS OF MRS DUBERLY by *Frances Isabella Duberly*—An Intrepid Victorian Lady's Experience of the Crimea and Indian Mutiny.

THE REBELLIOUS DUCHESS by *Paul F. S. Dermoncourt*—The Adventures of the Duchess of Berri and Her Attempt to Overthrow French Monarchy.

LADIES OF WATERLOO by *Charlotte A. Eaton, Magdalene de Lancey & Juana Smith*—The Experiences of Three Women During the Campaign of 1815: Waterloo Days by Charlotte A. Eaton, A Week at Waterloo by Magdalene de Lancey & Juana's Story by Juana Smith.

TWO YEARS BEFORE THE MAST by *Richard Henry Dana. Jr.*—The account of one young man's experiences serving on board a sailing brig—the Penelope—bound for California, between the years 1834-36.

A SAILOR OF KING GEORGE by *Frederick Hoffman*—From Midshipman to Captain—Recollections of War at Sea in the Napoleonic Age 1793-1815.

LORDS OF THE SEA by *A. T. Mahan*—Great Captains of the Royal Navy During the Age of Sail.

COGGESHALL'S VOYAGES: VOLUME 1 by *George Coggeshall*—The Recollections of an American Schooner Captain.

COGGESHALL'S VOYAGES: VOLUME 2 by *George Coggeshall*—The Recollections of an American Schooner Captain.

TWILIGHT OF EMPIRE by *Sir Thomas Ussher & Sir George Cockburn*—Two accounts of Napoleon's Journeys in Exile to Elba and St. Helena: Narrative of Events by Sir Thomas Ussher & Napoleon's Last Voyage: Extract of a diary by Sir George Cockburn.

AVAILABLE ONLINE AT **www.leonaur.com**
AND FROM ALL GOOD BOOK STORES

ALSO FROM LEONAUR
AVAILABLE IN SOFTCOVER OR HARDCOVER WITH DUST JACKET

ESCAPE FROM THE FRENCH by *Edward Boys*—A Young Royal Navy Midshipman's Adventures During the Napoleonic War.

THE VOYAGE OF H.M.S. PANDORA by *Edward Edwards R. N. & George Hamilton, edited by Basil Thomson*—In Pursuit of the Mutineers of the Bounty in the South Seas—1790-1791.

MEDUSA by *J. B. Henry Savigny and Alexander Correard and Charlotte-Adélaïde Dard*—Narrative of a Voyage to Senegal in 1816 & The Sufferings of the Picard Family After the Shipwreck of the Medusa.

THE SEA WAR OF 1812 VOLUME 1 by *A. T. Mahan*—A History of the Maritime Conflict.

THE SEA WAR OF 1812 VOLUME 2 by *A. T. Mahan*—A History of the Maritime Conflict.

WETHERELL OF H. M. S. HUSSAR by *John Wetherell*—The Recollections of an Ordinary Seaman of the Royal Navy During the Napoleonic Wars.

THE NAVAL BRIGADE IN NATAL by *C. R. N. Burne*—With the Guns of H. M. S. Terrible & H. M. S. Tartar during the Boer War 1899-1900.

THE VOYAGE OF H. M. S. BOUNTY by *William Bligh*—The True Story of an 18th Century Voyage of Exploration and Mutiny.

SHIPWRECK! by *William Gilly*—The Royal Navy's Disasters at Sea 1793-1849.

KING'S CUTTERS AND SMUGGLERS: 1700-1855 by *E. Keble Chatterton*—A unique period of maritime history-from the beginning of the eighteenth to the middle of the nineteenth century when British seamen risked all to smuggle valuable goods from wool to tea and spirits from and to the Continent.

CONFEDERATE BLOCKADE RUNNER by *John Wilkinson*—The Personal Recollections of an Officer of the Confederate Navy.

NAVAL BATTLES OF THE NAPOLEONIC WARS by *W. H. Fitchett*—Cape St. Vincent, the Nile, Cadiz, Copenhagen, Trafalgar & Others.

PRISONERS OF THE RED DESERT by *R. S. Gwatkin-Williams*—The Adventures of the Crew of the Tara During the First World War.

U-BOAT WAR 1914-1918 by *James B. Connolly/Karl von Schenk*—Two Contrasting Accounts from Both Sides of the Conflict at Sea During the Great War.

AVAILABLE ONLINE AT **www.leonaur.com**
AND FROM ALL GOOD BOOK STORES

www.ingramcontent.com/pod-product-compliance
Lightning Source LLC
Chambersburg PA
CBHW030228170426
43201CB00006B/152